Oh Calculus
A Workbook for MATH 221

Georgia B. Pyrros
University of Delaware

Kendall Hunt
publishing company

This edition has been printed directly from print-ready copy.

www.kendallhunt.com
Send all inquiries to:
4050 Westmark Drive
Dubuque, IA 52004-1840

Printed in the United States of America
10 9 8 7 6 5 4 3 2

Contents

Contents

Part 1

Limits, Continuity, and Derivatives

1.1 Limits

A Geometric Approach

Case 1

Let $f(x)$ be a function whose graph is shown below:

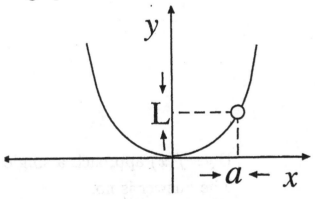

Does $f(x)$ approach some specific value as x approaches a?

The answer is yes.

As x values get closer and closer to a from both sides (how close? as close as we want.) the y values get closer and closer

1

to a single value L.

We call L the limit of $f(x)$ as x approaches a.
The notation is: $\boxed{\lim_{x \to a} f(x) = L}$

Note that in the preceding function $f(a)$ does not exist. When $x = a$ the y coordinate is not on the parabola. But, the limit of $f(x)$ as x approaches a does exist and is L.

So the value of the function and the limit of the function are two different things.

> **Definition:** If $f(x)$ approaches a single number L as x approaches a (but not equal to a), then L is the limit of $f(x)$ as x approaches a. $\boxed{\lim_{x \to a} f(x) = L}$

Keep in mind that when we say <u>approach</u> we do not mean <u>equal</u>.

Case 2

Let $f(x)$ be defined as follows:

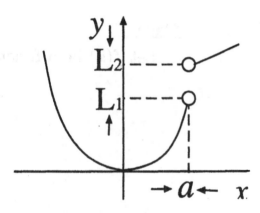

Does $f(x)$ approach a single number as x approaches a?
The answer is no.

Of course there are two ways to approach a, from the left and from the right.

As x values get closer and closer to a from the left, the y values get closer and closer to a single number L_1. We call L_1 the left limit of $f(x)$ as x approaches a. The notation is $\boxed{\lim_{x \to a^-} f(x) = L_1}$ (from the left)

As x values get closer and closer to a from the right, the y values get closer and closer to a single number L_2. We call L_2 the right limit of $f(x)$ as x approaches a (from the right).

The notation is $\boxed{\lim_{x \to a^+} f(x) = L_2}$

It is obvious in this case that as x values get closer and closer to a from both sides the y values do not get close to a single value (Since the left limit is L_1 and the right limit is L_2). Therefore, the limit of $f(x)$ as x approaches a does not exist.

Case 3

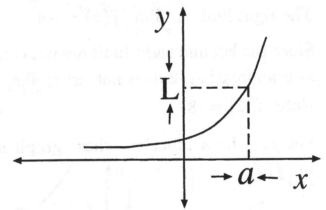

Does $f(x)$ approach a single value as x approaches a?

The answer is yes.

As x values get closer and closer to a from the left and from the right the y values get closer and closer to L. In this case, when $x = a$ the y coordinate $f(a)$ is L. So the value of the function at $x = a$ and the limit of the function as x approaches a are exactly the same: $\boxed{\lim_{x \to a} f(x) = f(a)}$

Examples

1. Let $f(x)$ be a function whose graph is shown. Find the $\lim\limits_{x \to -2} f(x)$.

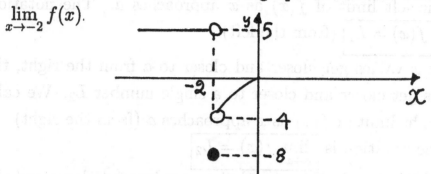

Obviously, this function consists of two pieces. So we must look at the left and right limits.

First, let us evaluate the left limit $\lim\limits_{x \to -2^-} f(x) = 5$

The right limit is $\lim\limits_{x \to -2^+} f(x) = -4$

Since the left and right limit are not equal, the limit of $f(x)$ as x approaches -2 does not exist; $\lim\limits_{x \to -2} f(x)$ does not exist.
Note: f(-2) = -8

2. Let $f(x)$ be a function whose graph is shown. Find the $\lim\limits_{x \to 4} f(x)$.

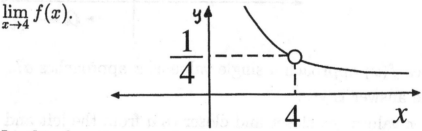

In this function a point is missing. $\lim\limits_{x \to 4^-} f(x) =$
$\lim\limits_{x \to 4^+} f(x) = \frac{1}{4}$

Since the left and right limit are equal, the $\lim\limits_{x \to 4} f(x)$ exists and equals $\frac{1}{4}$.

3.

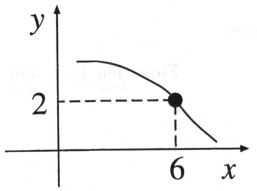

Find: $\lim\limits_{x \to 6} f(x)$

$$\lim_{x \to 6^-} f(x) = \lim_{x \to 6^+} f(x) = \lim_{x \to 6} f(x) = 2$$

4.

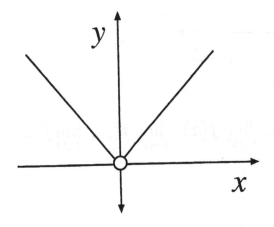

Find: $\lim\limits_{x \to 0} f(x)$

$$\lim_{x \to 0^-} f(x) = \lim_{x \to 0^+} f(x) = \lim_{x \to 0} f(x) = 0$$

Practice Exercises

1.

Find: $\lim\limits_{x \to 3^-} f(x)$, $\lim\limits_{x \to 3^+} f(x)$, $\lim\limits_{x \to 3} f(x)$

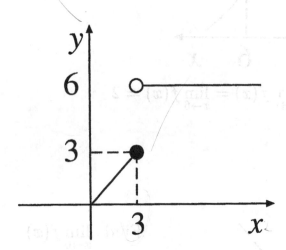

2.

Find: $\lim\limits_{x \to 1^-} f(x)$, $\lim\limits_{x \to 1^+} f(x)$, $\lim\limits_{x \to 1} f(x)$. Find $f(1)$.

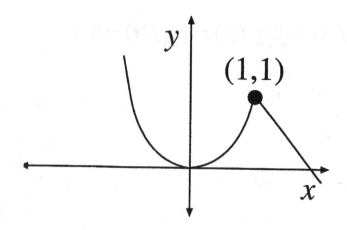

3.   Find: $\lim\limits_{x \to -1^-} f(x)$, $\lim\limits_{x \to -1^+} (x)$, $\lim\limits_{x \to -1} f(x)$. Find: $f(-1)$.

½   ½   ½   2

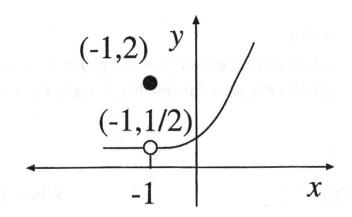

(-1,2)

(-1,1/2)

-1

4.   Find: $\lim\limits_{x \to 0^-} f(x)$, $\lim\limits_{x \to 0^+} f(x)$, $\lim\limits_{x \to 0} f(x)$. Find $f(0)$.

6   0   does not exist   0

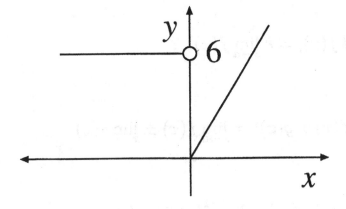

6

An Algebraic Approach

To evaluate limits algebraically we must use the rules of the limits.

Limit Rules

Let a, k, r be real numbers and $\lim_{x \to a} f(x)$, $\lim_{x \to a} g(x)$ exist. The following rules are true for left limit, right limit and limit.

1. $\lim_{x \to a} k = k$ Where k is a constant

2. $\lim_{x \to a} x = a$

3. $\lim_{x \to a} [k f(x)] = k \lim_{x \to a} f(x)$

4. $\lim_{x \to a} [f(x) \pm g(x)] = \lim_{x \to a} f(x) \pm \lim_{x \to a} g(x)$

5. $\lim_{x \to a} [f(x) \cdot g(x)] = \lim_{x \to a} f(x) \lim_{x \to a} g(x)$

6. $\lim_{x \to a} \left[\dfrac{f(x)}{g(x)} \right] = \dfrac{\lim_{x \to a} f(x)}{\lim_{x \to a} g(x)}$ as long as $\lim_{x \to a} g(x) \neq 0$

7. $\lim_{x \to a} [f(x)]^r = \left[\lim_{x \to a} f(x)\right]^r, r = $ positive integer

8. $\lim_{x \to a} \sqrt[r]{f(x)} = \sqrt[r]{\lim_{x \to a} f(x)}, r = $ positive integer

If r is even $\lim_{x \to a} f(x)$ must be positive.

The above rules can be combined to compute the limits of polynomial, rational, exponential, logarithmic, piece wise functions.

Examples

A. Polynomial Functions

1. Let $f(x) = 91$. Evaluate: $\lim_{x \to -3} f(x)$.

   Solution: $f(x)$ is a constant function (polynomial function that does not depend on x.)

   $$\lim_{x \to -3} f(x) = \lim_{x \to -3} 91 = 91$$

2. Evaluate: $\lim_{x \to -1} (x^3 + 3x^2 - 4x + 12)$

   Solution:

   $$\lim_{x \to -1} (x^3 + 3x^2 - 4x + 12) = \lim_{x \to -1} x^3 + \lim_{x \to -1} 3x^2 - \lim_{x \to -1} 4x + \lim_{x \to -1} 12$$

   $$= (-1)^3 + 3(-1)^2 - 4(-1) + 12 = 18$$

We evaluate the limit of a polynomial function by substitution.

Practice Exercises

1. Evaluate: $\lim\limits_{x \to 3}(x+3)(x-2)$

$\lim\limits_{x \to 3} ((3)+3)((3)-2)$
$(6)(1)$

$= \boxed{6}$

2. Evaluate: $\lim\limits_{h \to 0}(3xh^3 - 4h^2 + 2hx + 3)$

$\lim\limits_{h \to 0}(3x(0)^3 - 4(0)^2 + 2(0)x + 3)$

$= \boxed{3}$

3. $\lim\limits_{\Delta x \to 0}[4(x + \Delta x) - 3x]$

$\lim\limits_{\Delta x \to 0}[4(x + (0)) - 3x]$

$= [4(x) - 3x]$

$= 4x - 3x$

$= \boxed{x}$

B. Rational Functions

Examples

1. Evaluate: $\lim\limits_{x \to 2}\left(\frac{x^2+1}{x-3}\right)$

<u>Solution:</u> $\lim\limits_{x \to 2}\left(\frac{x^2+1}{x-3}\right) = \frac{\lim\limits_{x \to 2}(x^2+1)}{\lim\limits_{x \to 2}(x-3)} = \frac{2^2+1}{2-3} = \frac{4+1}{-1} = -5$

Since $\lim\limits_{x \to 2}(x-3) = -1 \neq 0$

> We evaluate the limit of a rational function by substitution as long as the limit of the denominator is different than zero.

2. Evaluate: $\lim\limits_{x \to 4}\left(\frac{x^2-16}{4-x}\right)$

<u>Solution:</u> In this case $\lim\limits_{x \to 4}(4-x) = 0$

So we can't apply rule 6. But, let's try to manipulate the expression algebraically.

For $x \neq 4$ $\frac{x^2-16}{4-x} = \frac{(x+4)(x-4)}{-(x-4)} = \frac{x+4}{-1} = -(x+4) = -x-4$

Therefore: $\lim\limits_{x \to 4}\left(\frac{x^2-16}{4-x}\right) = \lim\limits_{x \to 4}(-x-4) = -8$

3. Evaluate: $\lim\limits_{x\to 9}\left(\frac{3-\sqrt{x}}{9-x}\right)$

Solution: First investigate the limit of the denominator
$\lim\limits_{x\to 9}(9-x)=0$

Therefore, we can't apply rule #6. Let's use algebraic manipulation to simplify the expression $\frac{3-\sqrt{x}}{9-x}$

Multiply numerator and denominator by
$3+\sqrt{x}$ (conjugate of $3-\sqrt{x}$)

$$\frac{3-\sqrt{x}}{9-x}\cdot\frac{3+\sqrt{x}}{3+\sqrt{x}}=\frac{3^2-x}{(9-x)(3+\sqrt{x})}=\frac{9-x}{(9-x)(3+\sqrt{x})}=\frac{1}{3+\sqrt{x}}$$

Then $\lim\limits_{x\to 9}\left(\frac{3-\sqrt{x}}{9-x}\right)=\lim\limits_{x\to 9}\left(\frac{1}{3+\sqrt{x}}\right)=\frac{1}{6}$

4. Evaluate $\lim\limits_{x\to 5}\left[\frac{x^2-4}{(x-5)(x-2)}\right]$

Solution: Investigate the limit of the denominator
$\lim\limits_{x\to 5}(x-5)(x-2)=0$

In this case algebraic manipulation doesn't help since:

$$\frac{x^2-4}{(x-5)(x-2)}=\frac{(x+2)(x-2)}{(x-5)(x-2)}=\frac{x+2}{x-5}$$

$\lim\limits_{x\to 5}(x-5)=0.$

Therefore, $\lim\limits_{x\to 5}\left[\frac{x^2-4}{(x-5)(x-2)}\right]$ doesn't exist.

Practice Exercises

1. Evaluate: $\displaystyle\lim_{x \to 0} \frac{\sqrt{x+3} - \sqrt{3}}{x}$

$$\frac{\sqrt{x+3} - \sqrt{3}}{x} \cdot \frac{\sqrt{x+3} + \sqrt{3}}{\sqrt{x+3} + \sqrt{3}} = \frac{(x+3) - 3}{x(\sqrt{x+3} + \sqrt{3})}$$

$$= \frac{x}{x(\sqrt{x+3} + \sqrt{3})}$$

$$= \frac{1}{\sqrt{x+3} + \sqrt{3}}$$

$$\to \lim_{x \to 0} \frac{1}{\displaystyle\lim_{x \to 0}(\sqrt{x+3} + \sqrt{3})}$$

$$= \frac{1}{2\sqrt{3}}$$

2. Evaluate: $\displaystyle\lim_{h \to 0} \frac{(x+h)^2 - x^2}{h}$

$$\frac{(x+h)^2 - x^2}{h}$$

3. $\displaystyle\lim_{h \to 0} \frac{\frac{1}{(x+h)^2} - \frac{1}{x^2}}{h}$

4. $\displaystyle\lim_{x \to 2^-} \frac{2-x}{\sqrt{4-x^2}}$

5. Evaluate: $\displaystyle \lim_{x \to 9} \frac{\sqrt{x^2 - 5x - 36}}{8 - 3x}$

6. Suppose $a \neq 1$. Then evaluate:

$$\lim_{h \to 0} \frac{\frac{a+h+1}{a+h-1} - \frac{a+1}{a-1}}{h}$$

C. Piecewise Functions

Examples

1. Let

$$f(x) = \begin{cases} x & \text{if } x \le 3 \\ 6 & \text{if } x > 2 \end{cases} \quad \begin{array}{l} \text{Find the left limit,} \\ \text{right limit, and limit} \\ \text{of } f(x) \text{ as } x \to 3 \end{array}$$

$$\left. \begin{array}{l} \lim_{x \to 3^-} f(x) = \lim_{x \to 3^-} x = 3 \\[2mm] \lim_{x \to 3^+} f(x) = \lim_{x \to 3^+} 6 = 6 \end{array} \right\} \begin{array}{l} 3 \ne 6 \\[2mm] \lim_{x \to 3} f(x) \text{ doesn't exist} \end{array}$$

2. Let

$$f(x) = \begin{cases} 5 & \text{if } x < 8 \\ 9 & \text{if } x \ge 8 \end{cases}$$

Evaluate: $\lim_{x \to 8} f(x)$

$$\lim_{x \to 8^-} f(x) = 5, \quad \lim_{x \to 8^+} f(x) = 9 \quad 5 \ne 9 \quad \lim_{x \to 8} f(x) = \text{ doesn't exist}$$

Practice Exercises

1. Let

$$f(x) = \begin{cases} 6x & \text{if } -3 < x < -2 \\ \frac{5}{2}x - 10 & \text{if } -2 < x < -1 \\ 12x^2 - 1 & \text{if } x > -1 \end{cases}$$

Find $\lim_{x \to -2} f(x)$, $\lim_{x \to -1} f(x)$

2. Let

$$f(x) = \begin{cases} \frac{x^2+2x-3}{x+3} & \text{if } x \neq -3 \\ -4 & \text{if } x = -3 \end{cases}$$

Find $\lim\limits_{x \to -3} f(x)$

3. Let

$$f(x) = \begin{cases} 2x^2 - 1 & \text{if } -3 \leq x < -2 \\ 4 & \text{if } x = -2 \\ 5x & \text{if } -2 < x \leq 0 \end{cases}$$

Evaluate $\lim\limits_{x \to -2} f(x)$.   Justify your answer.

4. Let $f(x) = \begin{cases} x + 4 & \text{if } x \leq -3 \\ 10 - x^2 & \text{if } x > -3 \end{cases}$

Evaluate: $\lim\limits_{x \to -3} f(x)$

Infinite Limits, Vertical Asymptotes

Let the graph below represent a function $f(x)$.

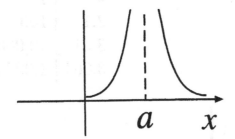

Does $f(x)$ approach some specific value as x approaches a?

The answer is no.

As x values get closer and closer to a from the left and from the right, the y values increase with no bound through positive numbers.

We say that the limit of $f(x)$ as x approaches a does not exist.

If $\lim_{x \to a} f(x)$ exists then the limit is a specific (finite) number.

If the values of $f(x)$ do not become closer and closer to a specific number as x gets closer and closer to a, then the limit does not exist.

Example 1

$$\text{Let } f(x) = \frac{1}{(x-3)^2}$$

Let us look at values of $f(x)$ as x gets close to 3.

From the left	
x	$f(x)$
2	1
2.9	100
2.99	10,000
2.999	1,000,000

From the right	
x	$f(x)$
4	1
3.1	100
3.01	10,000
3.001	1,000,000

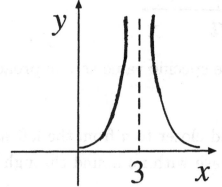

$$y = f(x) = \frac{1}{(x-3)^2}$$

It is obvious that as x gets close to 3 from the left and from the right, the values of $f(x)$ increase with no bound.

Mathematically we can write:

$$\lim_{x \to 3^-} f(x) = +\infty \text{ and } \lim_{x \to 3^+} f(x) = +\infty$$

We also write:

$$\lim_{x \to 3} f(x) = \infty \ , \ \lim_{x \to 3} \frac{1}{(x-3)^2} = \infty$$

This means that the limit does not exist.

Example 2

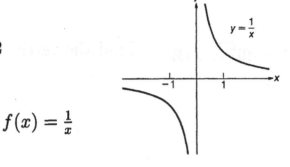

$$f(x) = \frac{1}{x}$$

The behavior of this function as $x \to 0$ can be described by the following:

$$\lim_{x \to 0^-} f(x) = -\infty \qquad \lim_{x \to 0^+} f(x) = +\infty$$

Vertical Asymptotes

In example 1 the vertical line $x = 3$ is a vertical asymptote. In example 2 the line $x = 0$ (y - axis) is a vertical asymptote. For a line $x = a$ to be a vertical asymptote of the graph of $f(x)$ any of the following must be true:

$$\left.\begin{array}{l} \lim_{x \to a} f(x) = \infty \\[1em] \lim_{x \to a} f(x) = -\infty \end{array}\right\} \qquad \left.\begin{array}{l} \lim_{x \to a^-} f(x) = \infty \\[1em] \lim_{x \to a^-} f(x) = -\infty \end{array}\right\} \qquad \begin{array}{l} \lim_{x \to a^+} f(x) = \infty \\[1em] \lim_{x \to a^+} f(x) = -\infty \end{array}$$

To find a vertical asymptote we set the denominator equal to zero. In other words, we find the points of discontinuity.

Practice Exercises

1. Let $f(x) = \frac{x+2}{x-1}$

   Use your graphic calculator to sketch the graph of $f(x)$.
   Find the vertical asymptote.

   Find $\qquad \lim_{x \to 1^-} f(x) \qquad , \qquad \lim_{x \to 1^+} f(x)$

2. Let $f(x) = \frac{x+2}{(x-1)(x-2)(x+3)}$.   Find the vertical asymptotes.

3. Let $f(x) = \frac{x^2}{x^2+3}$.   Find the vertical asymptotes.

Limits at Infinity. Horizontal Asymptotes

So far we have evaluated limits of the form $\lim_{x \to a} f(x)$. In other words, we have investigated the behavior of the function "near" to a point "a" where "a" is a real number.

Sometimes (and especially when we deal with physical applications), we need to know how the function behaves for very large or very small values of x, i.e. we need to evaluate limits of the form:

$$\lim_{x \to +\infty} f(x) \quad (x \text{ very large}) \qquad \lim_{x \to -\infty} f(x) \ (x \text{ very small})$$

The expression that we use for $x \to +\infty$ means that x moves with no bound through positive numbers, and for $x \to -\infty$ that x moves with no bound through negative numbers . To find the horizontal asymptote, we find the limit at infinity.

Let us first consider the simple example: to find

$$\lim_{x \to +\infty} \tfrac{1}{x}$$

Take : $x = 10$ then$\tfrac{1}{x} = $ $\tfrac{1}{10} = .1$

$x = 100$ then$\tfrac{1}{x} = $ $\tfrac{1}{100} = .01$

$x = 1000$ then$\tfrac{1}{x}$ $\tfrac{1}{1000} = .001$

$x = 10,000$ then$\tfrac{1}{x} = $ $\tfrac{1}{10,000} = 10^{-4}$

$x = 1,000,000$ then$\tfrac{1}{x} = $ $\tfrac{1}{1,000,000} = 10^{-6}$

and so on.

So we see that as x is getting larger and larger, $\tfrac{1}{x}$ is getting closer and closer to 0. So we conclude that

$$\lim_{x \to +\infty} \frac{1}{x} = 0.$$

Consider now another example. Find:

$$\lim_{x \to \infty} \frac{1}{x^2}$$

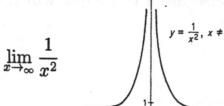

$y = \frac{1}{x^2}, x \neq 0$

Similarly, as before, take:

$$\lim_{x \to +\infty} \frac{1}{x^2}$$

Take : $x = 10$ then $\frac{1}{x^2} =$ $\frac{1}{10^2} = \frac{1}{100} = .01$

$x = 100$ then $\frac{1}{x^2} =$ $\frac{1}{(100)^2} = \frac{1}{10,000} = .0001$

$x = 1000$ then $\frac{1}{1,000,000} = 10^{-6}$

Again, as before, we have that:

$$\lim_{x \to +\infty} \frac{1}{x^2} = 0$$

The difference between these two examples is that $\frac{1}{x^2}$ approaches zero "faster" than $\frac{1}{x}$.

The above discussion leads to the following:

For every positive number r

$$\boxed{\lim_{x \to \pm\infty} \frac{1}{x^r} = 0}$$

Evaluation of Limits at Infinity

The limit rules we discussed before apply for the limits at infinity.

We also use a technique which is illustrated in the following example.

Example

Evaluate: $\lim\limits_{x\to\infty} \frac{5x^3-2}{4x^2-3x+1}$

<u>Solution:</u> Divide numerator and denominator by the highest power of x.

$$\lim_{x\to\infty} \frac{5x^3 - 2}{4x^2 - 3x + 1} = \lim_{x\to\infty} \frac{\frac{5x^3-2}{x^3}}{\frac{4x^2-3x+1}{x^3}}$$

$$= \lim_{x\to\infty} \frac{\frac{5x^3}{x^3}-\frac{2}{x^3}}{\frac{4x^2}{x^3}-\frac{3x}{x^3}+\frac{1}{x^3}}$$

$$= \lim_{x\to\infty} \frac{5-\frac{2}{x^3}}{\frac{4}{x}-\frac{3}{x^2}+\frac{1}{x^3}}$$

$$= \frac{5-0}{0-0+0} = \infty = \text{doesn't exist}$$

Practice Exercises

1. Find $\lim\limits_{x\to\infty} \frac{5x^2-3}{3x^3-2x^2-1}$

2. Find $\lim\limits_{x\to\infty} \frac{4x^4-3x^3-5x-1}{-5x^4-2x^2-3}$

3. Find $\lim\limits_{x \to \infty} \frac{3x^2 - 3x - 1}{x^3}$

4. Find the horizontal asymptote for $f(x) = \frac{x^2}{x^2 + 1}$

5. Find the horizontal asymptote for $f(x) = \frac{10}{16 - x^2}$

6. Find $\lim\limits_{x \to \infty} \frac{\sqrt{x^2 - 9}}{x}$

1.2 Continuity

A function $f(x)$ is continuous at $x = a$ (a any real number) if the value of the function at $x = a$ is the same as the limit of $f(x)$ as x approaches a.

That is $\boxed{f(a) = \lim_{x \to a} f(x)}$

This definition can be broken into three steps.

Step 1. $f(x)$ must be defined at $x = a$

Step 2. the $\lim_{x \to a} f(x)$ must exist

Step 3. $f(a) = \lim_{x \to a} f(x)$

Examples

1. Let $f(x) = x^3 + 1$ (polynomial function). Is $f(x)$ continuous at $x = 2$?
   Solution:

   Step 1. The value fo the function at $x = 2$ is:

   $$f(2) = 2^3 + 1 = 9$$

   Step 2. $\lim_{x \to 2} f(x) = \lim_{x \to 2} (x^3 + 1) = 9$.
   Since $f(2) = \lim_{x \to 2} (x^3 + 1) = 9$, the function $x^3 + 1$ is continuous at $x = 2$.
   Note: Polynomial functions are continuous functions.

2. Let $f(x) = \frac{x^2-9}{x+3}$ (rational function). Is $f(x)$ continuous at $x = -3$?

Solution:

Step 1. $f(-3)$ does not exist

Step 2. $\lim\limits_{x \to -3} \frac{x^2-9}{x+3} = \lim\limits_{x \to -3} \frac{(x-3)(x+3)}{x+3} = \lim\limits_{x \to -3} (x - 3) = -6.$

The $\lim\limits_{x \to -3} f(x)$ exists but the function is not defined at $x = -3$. Therefore $f(x)$ is discontinuous at $x = -3$.

The graph is:

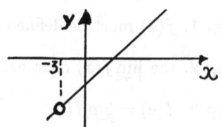

Note: $x = -3$ is the only point of discontinuity for $f(x)$.

3. Let $f(x) = \begin{cases} 1 - 2x & \text{if } -1 < x < 0 \\ 1 & \text{if } x = 0 \\ x^2 + 1 & \text{if } 0 < x < 2 \end{cases}$

Is $f(x)$ continuous at $x = 0$?

Solution:

Step 1: $f(0) = 1$

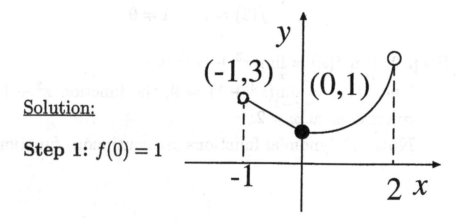

Step 2: We have to evaluate the left limit and the right limit.

$$\lim_{x \to 0^-} f(x) = \lim_{x \to 0^-} (1 - 2x) = 1$$

$$\lim_{x \to 0^+} f(x) = \lim_{x \to 0^+} (x^2 + 1) = 1.$$

Therefore the $\lim_{x \to 0} f(x) = 1$. Since $f(0) = \lim_{x \to 0} f(x) = 1$ the function is continuous at $x = 0$.

Practice Exercises

1. Let $f(x) = \begin{cases} 3x - 4 & \text{if } -1 < x \leq 1 \\ \\ 2x^2 - 3 & \text{if } 1 < x < 2 \end{cases}$

Is $f(x)$ continuous at $x = 1$?

2. Let $f(x) = \begin{cases} \frac{x^2 - 25}{x+5} & \text{if } x \neq 5 \\ \\ -4 & \text{if } x = 5 \end{cases}$

Is $f(x)$ continuous at $x = 5$?

3. Find c such that the function

$$f(x) = \begin{cases} 5 - 2x & \text{if } -5 < x < -4 \\ \\ 2c & \text{if } x = -4 \\ \\ \frac{-13}{4}x & \text{if } -4 < x < 2 \end{cases}$$

is continuous at $x = -4$.

1.3 Derivatives

The Secant Line

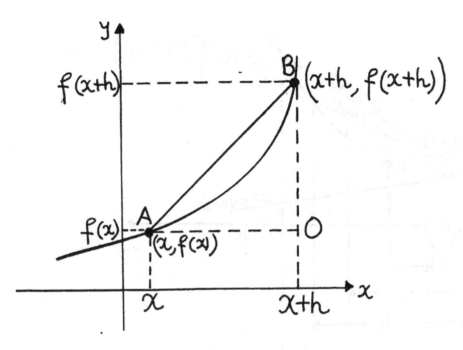

Let $f(x)$ be a function whose graph is shown and $(x, f(x))$, $(x + h, \ f(x + h))$ two points on the graph of $f(x)$. Then the line that joins these two points is called the secant line to the graph. The slope of the secant line is:

slope of the secant line =

$$m_{sec} = \frac{OB}{OA} = \frac{rise}{run} = \frac{f(x + h) - f(x)}{h} = \frac{\Delta y}{\Delta x}$$

The slope of the secant line is also called the average rate of change or the difference quotient.

The Tangent Line

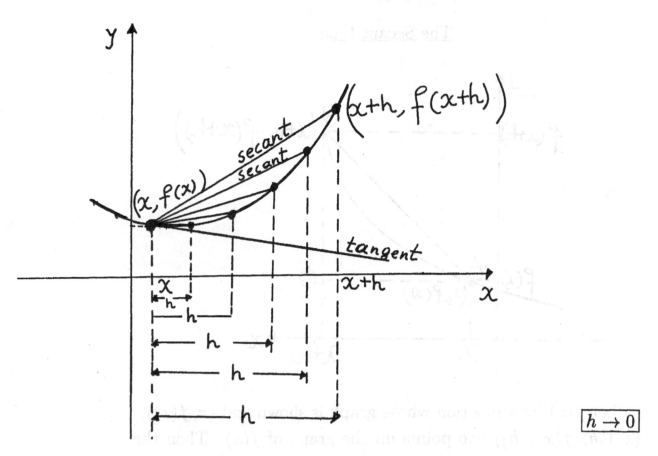

Let $f(x)$ be a function whose graph is shown. Several secant lines are also shown. The point $(x, f(x))$ on the graph of $f(x)$ stays fixed. As h becomes smaller and smaller $(h \to 0)$ new secant lines are created and eventually we realize that the <u>limiting position</u> of the secant line is the tangent line to the graph of $f(x)$ at the point $(x, f(x))$. The slope of the tangent line is:

$$\text{slope of the tangent line} = m_{\tan}(x) = \lim_{h \to 0} m_{\sec}$$

But we know that the slope of the secant line is:

$$m_{\text{sec}} = \frac{f(x+h) - f(x)}{h}$$

Therefore

$$m_{\text{tan}}(x) = \lim_{h \to 0} \frac{f(x+h) - f(x)}{h}$$

The slope of the tangent line to the graph at the point $(x, f(x))$ is called the derivative of $f(x)$ at that point. The notation is $f'(x)$ (prime notation) or $\frac{df(x)}{dx}$ or $\frac{dy}{dx}$ (Leibniz notation). Therefore the definition of the derivative is:

$$\boxed{f'(x) = \frac{df(x)}{dx} = \lim_{h \to 0} \frac{f(x+h) - f(x)}{h}}$$

or we can write:

$$\boxed{\frac{dy}{dx} = \lim_{\Delta x \to 0} \frac{\Delta y}{\Delta x}}$$

The derivative is also called the instantaneous rate of change or simply the rate of change.

Examples

1. Let us now consider a specific function and find the derivative by using the definition. Let $f(x) = \frac{1}{x}$. Find $f'(x)$. The definition of $f'(x)$ is:

$$\boxed{f'(x) = \lim_{h \to 0} \frac{f(x+h) - f(x)}{h}}$$

First we create the difference quotient (or the slope of the secant line)

$$\frac{f(x+h)-f(x)}{h} = \frac{\frac{1}{x+h}-\frac{1}{x}}{h}$$

Next we simplify the difference quotient

$$\frac{\frac{1}{x+h}-\frac{1}{x}}{h} = \frac{\frac{x-x-h}{(x+h)x}}{h} = \frac{-h}{(x+h)x \cdot h} = \frac{-1}{(x+h)x}.$$

(review complex fractions)

Now we are ready to calculate the limit of the difference quotient.

$$\lim_{h\to 0}\frac{-1}{(x+h)x} = \frac{-1}{(x+0)x} = \frac{-1}{x^2}$$

Therefore the derivative of $f(x)=\frac{1}{x}$ is: $f'(x)=-\frac{1}{x^2}$.

2. Let $f(x)=\sqrt{x}$. Find $f'(4)$ by using the limit definition of the derivative.

First we find $f'(x)$.

$$f'(x) = \lim_{h\to 0}\frac{f(x+h)-f(x)}{h} = \lim_{h\to 0}\frac{\sqrt{x+h}-\sqrt{x}}{h} =$$

$$\lim_{h\to 0}\frac{\sqrt{x+h}-\sqrt{x}}{h}\cdot\frac{\sqrt{x+h}+\sqrt{x}}{\sqrt{x+h}+\sqrt{x}} = \lim_{h\to 0}\frac{x+h-x}{h(\sqrt{x+h}+\sqrt{x})} =$$

$$\lim_{h\to 0}\frac{1}{\sqrt{x+h}+\sqrt{x}} = \frac{1}{\sqrt{x}+\sqrt{x}} = \frac{1}{2\sqrt{x}}.$$

$$f'(x) = \frac{1}{2\sqrt{x}}$$

$$f'(4) = \frac{1}{2\cdot 2} = \frac{1}{4}$$

Ques. How do we find limits algebraically?

Ans. By substitution if we are allowed.

Practice Exercises

1. Let $f(x) = \frac{1}{x-3}$. Create the difference quotient for the points: $(x, f(x))$, $(x + \Delta x, f(x + \Delta x))$. Then simplify.

$$\frac{f(x+h) - f(x)}{h} = \frac{\frac{1}{(x+h)-3} - \frac{1}{x-3}}{h} = \frac{h}{(x+h)}$$

2. Let $f(x) = \sqrt[3]{x}$. Create the difference quotient for the points: $(1, 1)$, $(8, 2)$.

3. Let $f(x) = \frac{1}{x^2}$. Use the limit definition of the derivative to find $\frac{df(x)}{dx}$.

4. Let $f(x) = \frac{1}{\sqrt{x-3}}$. Use the definition of the derivative to find $f'(x)$.

5. Let $f(x) = \frac{x-1}{x+2}$. Create the slope of the secant line for the points: $(x, f(x))$, $(x+h, f(x+h))$.

6. Let $f(x) = \sqrt{x^2+1}$. Use the definition of the derivative to find $\frac{df(x)}{dx}\Big|_{x=1}$.

1.4   Differentiation Rules

The process of finding the derivative of a function is called differentiation. Since it is often time consuming to find the derivative from the limit definition, rules have been developed for finding derivatives quickly and efficiently.

1. $\frac{d}{dx}c = 0$ (derivative of a constant is zero.)

2. constant multiple rule

$$\frac{d}{dx}[cf(x)] = c\frac{df(x)}{dx}$$

3. $\frac{d}{dx}[f(x) \pm g(x)] = \frac{d}{dx}f(x) \pm \frac{d}{dx}g(x)$

Differentiation Rule for Power Functions (<u>Power Rule</u>)

Let $\boxed{f(x) = x^r}$ r any real number (if $r < 0$ we assume that

$x \neq 0$) then $\boxed{f'(x) = \frac{df(x)}{dx} = rx^{r-1}}$

Examples

1. Find $f'(x)$ if $f(x) = 4x^3 + \frac{7}{2}x^2$

$$f'(x) = \frac{df(x)}{dx} = 12x^2 + \frac{7}{2} \cdot 2 \cdot x = 12x^2 + 7x$$

2. Find $\frac{df(x)}{dx}$ if $f(x) = \frac{2}{x^3} - \frac{1}{x^2}$.
   Rewrite $f(x)$ as : $f(x) = 2x^{-3} - x^{-2}$

$$\frac{df(x)}{dx} = (-3) \cdot 2x^{-4} - (-2)x^{-3} = -6x^{-4} + 2x^{-3} = \frac{-6}{x^4} + \frac{2}{x^3}$$

3. Let $f(x) = \sqrt[3]{x} + \frac{4}{\sqrt{x^3}}$. Find $f'(x)$.

$$f(x) = x^{\frac{1}{3}} + 4x^{-3/2}$$

$$f'(x) = \frac{1}{3}x^{\frac{1}{3}-1} + 4(-\frac{3}{2})x^{-\frac{3}{2}-1}$$

$$\boxed{f'(x) = \frac{1}{3}x^{-\frac{2}{3}} - 6x^{-\frac{5}{2}}}$$

General Power Rule

Differentiation Rule for power functions that are composite functions.

$$\boxed{\frac{d}{dx}[f(x)]^r = rf(x)^{r-1} \cdot f'(x)}$$

Examples

1. Let $f(x) = (2x + 3)^{4/3}$. Find $f'(x)$

$$\frac{df(x)}{dx} = f'(x) = \frac{4}{3}(2x+3)^{\frac{4}{3}-1} \cdot (2x+3)'$$

$$\frac{df(x)}{dx} = \frac{4}{3}(2x+3)^{\frac{1}{3}} \cdot 2$$

$$\frac{df(x)}{dx} = \frac{8}{3}(2x+3)^{\frac{1}{3}} = \frac{8\sqrt[3]{2x+3}}{3}$$

Practice Exercises

1. Let $f(x) = \frac{1}{\sqrt{x}} - \frac{1}{\sqrt[3]{x}} + 4x^2$. Differentiate $f(x)$.

2. Let $f(x) = \frac{1}{\sqrt{x^2-4}}$. Evaluate $f'(4)$.

3. Let $f(x) = \frac{4}{2(x^2+7)^3}$. Evaluate $\frac{df(x)}{dx}\Big|_{x=1}$

4. Let $f(x) = \frac{x^2+2x+1}{4}$. Find $f'(x)$.

5. Let $f(x) = (x^4 + 4x)^3$. Find $f'(x)$.

6. Let $f(t) = \sqrt{2t - 1} + \sqrt{5t + 1}$. Find $\frac{df(t)}{dt}$.

1.5 Continuity and Differentiability

If a function $f(x)$ is differentiable at $x = a$ then $f(x)$ is continuous at $x = a$. In other words: differentiability guarantees continuity. The opposite is not true.
Continuity does not guarantee differentiability.

To clear this concept we must look at the definition of the derivative. If $f(x)$ is continuous at $(x, f(x))$, then the $\lim\limits_{h \to 0} \frac{f(x+h)-f(x)}{h}$ must exist for the function $f(x)$ to be differentiable at $(x, f(x))$.

Example

Let $f(x) = \begin{cases} x^2 & \text{if } x \le 0 \\[2mm] x & \text{if } x \ge 0 \end{cases}$

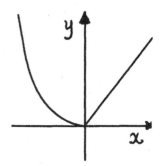

Is $f(x)$ continuous and differentiable at $x = 0$?
$\underline{f(x) \text{ is continuous at } x = 0.}$

Now let us look at the definition of the derivative. We know that $f'(x) = \lim\limits_{h \to 0} \frac{f(x+h)-f(x)}{h}$. For the derivative to exist the limit of the difference quotient $\frac{f(x+h)-f(x)}{h}$ $\underline{\text{must exit}}$.

For the given function we have:

$$\lim_{h \to 0^-} \frac{f(x+h)-f(x)}{h} = \lim_{h \to 0^-} \frac{(x+h)^2 - x^2}{h} = \lim_{h \to 0^-} (2x+h) = 2x$$

$$\lim_{h \to 0^+} \frac{f(x+h)-f(x)}{h} = \lim_{h \to 0^+} \frac{x+h-x}{h} = 1$$

$\boxed{\text{At } x = 0}$

the left limit of the difference quotient $\quad = 0$
and the right limit of the difference quotient $\ = 1$

Therefore the limit of the difference quotient does not exist.

Therefore $f'(x)$ does not exist and the given function $f(x)$ is not differentiable at $x = 0$. We say that the graph of $f(x)$ has a corner at $x = 0$.

Geometrically, the tangent line to the graph at $x = 0$ appears to be two different lines. So it is like having a left derivative and a right derivative. Functions that have a corner at $x = a$ are not differentiable at $x = a$.

Practice Exercises

1. Left $f(x) = |x|$. Is $f(x)$ continuous and differentiable at $x = 0$? Justify your answer.

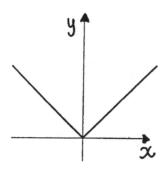

2. Let $f(x) = \begin{cases} x^2 & \text{if } -\infty < x \le 1 \\ 1 & \text{if } 1 \le x < \infty \end{cases}$

   Is $f(x)$ continuous and differentiable at $x = 1$?

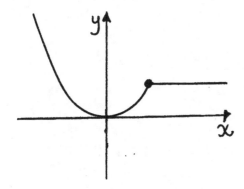

3. Investigate the continuity and differentiability of the function

$$f(x) = 2 + |x| \quad \text{in } (-\infty, \infty)$$

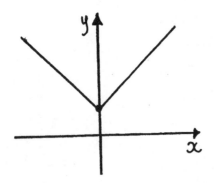

4. Let $f(x) = |\sqrt[3]{x}|$. Is $f(x)$ continuous and differentiable at $x = 0$? Explain.

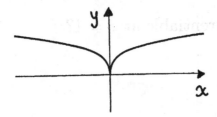

5. Let $f(x) = \sqrt[3]{x}$. Find the points of discontinuity and the points of non-differentiability.

6. Let $f(x) = \begin{cases} \frac{x^2+4x+3}{x+3} & \text{if } x \neq -3 \\ \\ 0 & \text{if } x = -3 \end{cases}$

Is $f(x)$ continuous and differentiable at $x = -3$?

1.6   The Derivative as a Rate of Change

Velocity and Acceleration

Assume that you are driving your car heading home. The rate of change of your distance with respect to time is called speed. If you also consider the direction in which the car is moving, the rate is called velocity. Let $s(t)$ be the distance in miles and t the time in hours.

If the distance between the university and your home is 150 miles and it takes you 3 hours to reach home, your average speed is:

$$\frac{\Delta s(t)}{\Delta t} = \frac{150 \text{ miles}}{3 \text{ hours}} = 50 \text{ miles/hour}$$

Of course the speedometer does not give you the average speed.

When you look at your speedometer you read your instantaneous speed, i.e., the rate of change of your distance with respect to time at <u>that instant</u>. How do we measure <u>that instant</u>? We take smaller and smaller time intervals. Therefore we calculate the limit of the average speed as $\Delta t \to 0$.

Let $s(t)$ be the position function of a moving object.

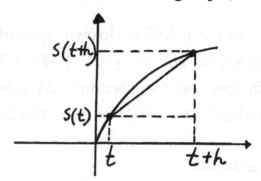

The <u>average velocity is:</u>

$$v_{\text{aver.}} = \frac{\Delta s(t)}{\Delta t}$$

or

$$\boxed{v_{\text{aver.}} = \frac{s(t+h) - s(t)}{h}}$$

The <u>instantaneous velocity</u> or simply velocity is:

$$v(t) = \frac{ds(t)}{dt} = s'(t)$$

or

$$\boxed{v(t) = \lim_{h \to 0} \frac{s(t+h) - s(t)}{t}}$$

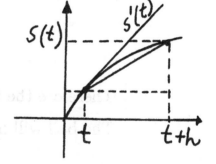

If velocity increases then the object accelerates. Then instant acceleration is

$$a(t) = \frac{dv(t)}{dt} = v'(t) = \frac{d^2 s(t)}{dt^2} = \lim_{h \to 0} \frac{v(t+h) - v(t)}{h}$$

Examples

1. Suppose a ball is thrown upward. Its position at time t is given by $s(t) = -16t^2 + 80t + 7$, where $s(t)$ is measured in feet and t in seconds. At what time does the ball stop rising? At what time does the ball hit the ground?

Solution:

$$s(t) = -16t^2 + 80t + 5$$

The ball will stop when $v(t) = 0$ since

$$v(t) = \frac{ds(t)}{dt} = s'(t)$$

$$v(t) = -32t + 80 = 0$$

$$-32t = -80$$

$$t = \tfrac{5}{2} \text{ sec}$$

Therefore the ball will stop rising after 2.5 sec.

The ball will hit the ground when

$$s(t) = 0$$

$$s(t) = -16t^2 + 80t + 5 = 0$$

by solving the quadratic function we find that the ball will hit the ground after 5.06 sec.

2. A rock is dropped from a helicopter. Its distance from the ground is given by

$$s(t) = -16t^2 + 32,000 \text{ feet}$$

What is the velocity of the rock when it hits the ground?

Solution:

$$v(t) = s'(t) = -32t$$

But the rock will hit the ground when

$$s(t) = 0, -16t^2 + 32,000 = 0$$

$$t^2 = 2000$$

$$t = 44.7 \text{ sec.}$$

$$v(44.7) = -32(44.7) = -1,430.4 \text{ ft/sec}$$

Marginal Cost. Marginal Profit

An important concept is that of marginal cost. Marginal cost of an item is the cost of producing that item. For example the marginal cost of producing the 15th item is the cost of producing 15 items minus the cost of producing 14 items.

If $C(x)$ is a cost function (x is the number of items produced), the derivative $C'(x) = \frac{dC(x)}{dx}$ is the marginal cost.

Therefore the marginal cost is:

$$C'(x) = \frac{dC(x)}{dx} = \lim_{h \to 0} \frac{C(x+h) - C(x)}{h}$$

The average cost for producing h items is

$$\frac{\Delta C(x)}{\Delta x} = \frac{C(x+h) - C(x)}{h}$$

Example

Let a cost function be $C(x) = 3.7 + 2\sqrt{x}$, where $C(x)$ is given in dollars and x is the number of items. Find the marginal cost of producing the 4th and 25th item.

Solution:

$$C(x) = 3.7 + 2\sqrt{2}$$

$$C'(x) = \frac{dC(x)}{dx} = 2\tfrac{1}{2}x^{-\frac{1}{2}} = \frac{1}{\sqrt{x}}$$

$$C'(4) = \tfrac{1}{2} = 0.5 = 50\cent$$

The cost of producing the 4th item is $50\cent$.

$$C'(25) = \frac{1}{5} = 20\cent$$

Practice Problems

1. Let $s(t) = 4t^3 + 8t^2$. Find the velocity and acceleration.

2. A ball is thrown upward from a 12 foot platform. Its position at time t seconds is given by $s(t) = 12 + 96t - 16t^2$ feet. Find the time required to reach the highest point, the distance that the highest point is above the ground, and the initial velocity.

3. If $s(t) = \frac{1}{3}t^3 - 3t^2 + 16t$, where the distance s is measured in centimeters and the time t in seconds, at what time will the velocity be 8 cm/sec?

4. The cost to produce x hand-made weather vanes is given by $C(x) = 100 + 8x - x^2 + 4x^3$. Find the marginal cost for $x = 6$.

5. A tumor has the shape of a cone. The tumor is growing along the height of the cone, the radius is fixed at 2 cm. Find the rate of change of the volume of the tumor with respect to height.

   Note: Volume of a cone: $V = \frac{1}{3}\pi r^2 \cdot h$

6. If the price of a product is given by $P(x) = \frac{-1000}{x} + 1000$, where x is the demand for the product, find the rate of change of price when the demand is $x = 10$.

Part 2

Applications of the Derivative

2.1 Curve Sketching

If $f'(x) > 0$ on an interval (a, b) then the graph of $f(x)$ is rising on that interval.

If $f'(x) < 0$ on an interval (a, b) then the graph of $f(x)$ is falling on that interval.

If $f''(x) > 0$ on an interval (a, b) then the graph of $f(x)$ is concave up on (a, b).

If $f''(x) < 0$ on an interval (a, b) then the graph of $f(x)$ is concave down on (a, b).

Sign of the Derivatives on an interval (a, b)	Graph of $f(x)$ on (a, b)
$f'(x) > 0$ $f''(x) > 0$	$f(x)$ is increasing with an increasing slope.
$f'(x) > 0$ $f''(x) < 0$	$f(x)$ is increasing with a decreasing slope
$f'(x) < 0$ $f''(x) > 0$	$f(x)$ is decreasing with increasing slope. (The slope becomes less negative.)
$f'(x) < 0$ $f''(x) < 0$	$f(x)$ is decreasing with decreasing slope. (The slope becomes more negative.)

The critical points of a function $f(x)$ are given by the roots and by the points of discontinuity of the first derivative. A relative extreme point can occur only at a critical point. When $f'(x) = 0$ the $m(x) = 0$ (slope is zero), therefore the curve of $f(x)$ has a horizontal tangent line.

Examples

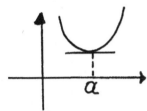

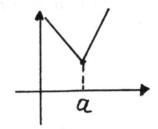

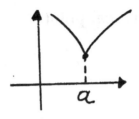

$f'(a) = 0$ $f'(a)$ doesn't exist $f'(a)$ doesn't exist

horizontal tangent line   vertical tangent line

At $x = a$ all $f(x)$ have a local minimum.

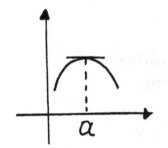

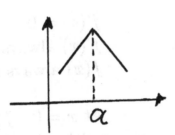

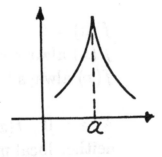

$f'(a) = 0$ $f'(a)$ doesn't exist $f'(a)$ doesn't exist

At $x = a$ all functions have a local maximum

The function does not necessarily have a relative extreme at a critical point.

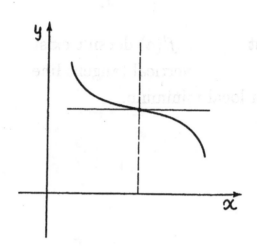
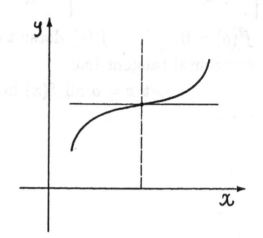

$f'(a) = 0$
$f'(x)$ always negative
$f(x)$ always falling

At $x = 0$ $f(x)$ has
neither local max
nor local min.

$f'(a) = 0$
$f'(x)$ always positive
$f(x)$ always rising

At $x = 0$ $f(x)$ has
neither local max
nor local min.

Example

Let $F(x) = 1 + x^{\frac{2}{3}}$. Use $f'(x)$ and $f''(x)$ as sketching tools for $f(x)$. Locate all maxima and minima.

$$f'(x) = \frac{2}{3}x^{-\frac{1}{3}} = \frac{2}{3\sqrt[3]{x}}$$

At $x = 0$ $f'(x)$ is not defined. $f'(x)$ has no roots. The critical points: $x = 0$

$f'(x)$

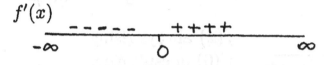

$f''(x) = -\frac{2}{9}x^{\frac{4}{3}} = \frac{-2}{9\sqrt[3]{x^4}}$

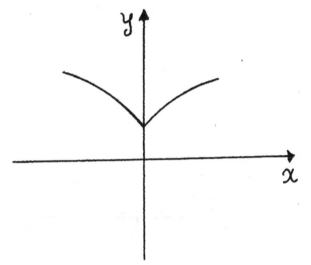

$f''(x)$

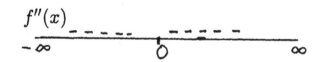

$f(x)$ always concave down

Points of Inflection

A point of inflection is a point on the curve of $f(x)$ at which the concavity changes from up to down or from down to up.

Example

Let $f(x) = x^{\frac{1}{3}}$. Find all local extrema, and inflection points

$$f'(x) = \frac{1}{3}x^{-\frac{2}{3}} = \frac{1}{3\sqrt[3]{x^2}} \qquad\qquad f''(x) = -\frac{2}{9}x^{-\frac{5}{3}} = \frac{-2}{9\sqrt[3]{x^5}}$$

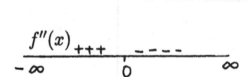

$f(x)$ always rising
$f'(0)$ doesn't exist,
    (vertical tangent line)

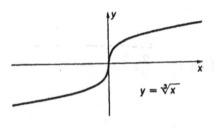

$y = \sqrt[3]{x}$

At $x = 0$ $f(x)$ has a point of inflection.

Exercises

1. Use your graphing calculator to sketch: $\frac{-1}{(x-2)^2}$ which is the first derivative of a function $f(x)$.

   Does $f(x)$ have a local max or min? Is $f(x)$ rising? Is $f(x)$ falling? Justify your answers.

$$f'(x) = \frac{-1}{(1-2)^2} \quad = \quad \frac{-1}{(3-2)^2}$$

f'(1)   2   f'(3)

$\ominus$ $\ominus$

$f(x)$ decreasing $(-\infty, 2)$ decreasing $(2, \infty)$

$f(x)$ is falling $f'(x) < 0$

no local max or min, vertical asymptote at $x = 2$

2. Use your graphic calculator to sketch $f(x) = 3x^5 - 5x^4 + 4$. What can you say about the sign of $\frac{d^2f(x)}{dx^2}$? Does the second derivative cross the x-axis? Justify your answers.

$f'(x) = 15x^4 - 20x^3$
$= 5x^3(3x - 4)$
$x = 0 \quad x = \frac{4}{3}$

$f''(x) = 60x^3 - 60x^2$
$= 60x^2(x - 1)$
$x = 0 \quad x = 1$

$f''(-1) \quad 0 \quad f''(\frac{1}{2}) \quad 1 \quad f''(2)$

$f''(x) = 60(-1)^2(-1-1) \quad = 60(\frac{1}{2})^2(\frac{1}{2}-1) \quad = 60(2)^2(2-1)$

$\ominus \qquad \ominus \qquad \oplus$

$f(x)$ concave down | concave down | concave up

$(-\infty, 0) \quad (0, 1) \quad (1, \infty)$

$\cap \qquad \cap \qquad \cup$

inflection pts. $(1, f(1))$
$= (1, 2)$

3. Let $\frac{d^2 f(x)}{dx^2} = \frac{-10}{(x+1)^3}$ Use your graphic calculator to sketch $\frac{d^2 f(x)}{dx^2} = \frac{-10}{(x+1)^3}$ (second derivative of the function $f(x)$). What can you say about $f(x)$? Does $f(x)$ have any inflection points? ~~What about the concavity of $f(x)$?~~ Justify your answers.

$$f''(x) = \frac{-10}{(x+1)^3} \qquad x = -1$$

$$f''(x) = \frac{-10}{(-2+1)^3} \qquad = \frac{-10}{(0+1)^3}$$

$\boxed{f''(-2)}$ concave up $(-\infty, -1)$ $\boxed{f''(0)}$ concave down $(-1, \infty)$

inflection pts.   Yes

4. Let $12x^2 - 24x$ be the second derivative of the function $f(x)$. Use your graphic calculator to sketch $f''(x)$. By looking at the graph of $f''(x)$ can you decide if $f(x)$ has any inflection points? Justify your answer.

f(x) has inflection points
b/c f''(x) is >0 and <0
=> changes in concavity in f(x)

5. Let $f(x) = 2x^3 + 3x^2 - 12x$. Sketch $f(x)$ (use your graphic calculator). What can you say about the sign of $f'(x)$?

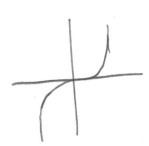

$$f'(x) = 6x^2 + 6x$$
$$6x(x+1)$$
$$x = 0 \qquad x = -1$$

always rising $f'(x) = 6(-1)(-1+1)' = 6(\tfrac{1}{2})(\tfrac{1}{2}+1) \; /= 6(2)(2+1)$

$f'(-1)$ 0 $f'(\tfrac{1}{2})$ 1 $f'(2)$

$\oplus$ $\oplus$ $\oplus$

$f(x)$ increasing , increasing , increasing

$(-\infty, 0)$ | $(0,1)$ | $(1, \infty)$

6. Let $f(x) = \frac{2x+1}{x-1}$. Find all horizontal and all vertical asymptotes. Sketch $f(x)$.

$$f(x) = \frac{2x+1}{x-1}$$

$$\lim_{x \to \infty} \frac{2x+1}{x-1} = 2 \quad \text{horizontal}$$

$$x = 1 \quad \text{vertical}$$

7. Use your graphic calculator to sketch the graph of $f(x) = x + \frac{1}{x} - 5$. Determine the asymptote of $f(x)$ (if it exists) as $|x|$ gets large.

8. Let $f(x) = \frac{x-3}{x^2+7}$. Find the vertical and horizontal asymptotes.

2.2 Optimization

We have 250 feet of fencing and we like to use it all to enclose the largest possible rectangular garden. Find the dimension of the garden.

The largest garden is the one with maximum area.

The garden might have the following shapes.

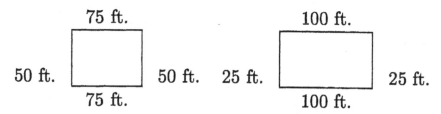

$$A = 75 \times 50 = 3750 \ (\text{ft.})^2 \qquad 100 \times 25 = 2500 \ (\text{ft.})^2$$

Many other shapes are possible. The area depends on the dimension of the rectangle.

<u>Solution:</u>

1. Let x be the length and y the width. Then

$$
\begin{aligned}
A &= xy \\
2x + 2y &= 250 \\
y &= 125 - x \\
A(x) &= x(125 - x) \\
A(x) &= -x^2 + 125x \\
A'(x) &= 2x + 125 = 0 \\
2x &= 125 \\
x &= 62.5 \ \text{ft.} \\
y &= 62.5 \ \text{ft.}
\end{aligned}
$$

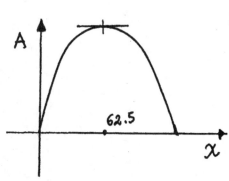

Practice Exercises

1. Find two positive numbers whose sum is 20 and whose product is as large as possible.

2. Prove that among all rectangles with given perimeter, the square has the largest area.

3. A closed container in the form of a right circular cylinder is to be made out of tin. For given surface area, what dimensions should the container have in order to maximize the volume?

4. A closed box is to be made out of heavy cardboard. If the base of the box is a rectangle twice as long as it is wide, and the box is to hold 9 cubic feet, what dimensions require the least amount of cardboard?

2.3   Techniques of Differentiation

$$\textbf{Product Rule:} [f(x)g(x)]' = f'(x)g(x) + g'(x)f(x)$$

$$\textbf{Quotient Rule:} \left[\frac{f(x)}{g(x)}\right]' = \frac{f'(x)g(x) - g'(x)f(x)}{g^2(x)}$$

Chain Rule (Derivative of a Composite Function)

$$\frac{d}{dx}f(g(x)) = f'(g(x)) \cdot g'(x)$$

$$\text{if } g(x) = u, \; f(u) = y$$

$$\frac{dy}{dx} = \frac{dy}{du} \cdot \frac{du}{dx}$$

Examples

1. Let $f(x) = \sqrt{x}(x^2 + 7)$. Differentiate.

   Solution:

$$\frac{df(x)}{dx} = \tfrac{1}{2}x^{-\frac{1}{2}}(x^2 + 7) + 2x\sqrt{x}$$

$$\frac{df(x)}{dx} = \frac{x^2 + 7}{2\sqrt{x}} + 2x\sqrt{x}$$

$$\frac{df(x)}{dx} = \frac{x^2 + 7 + 4x^2}{2\sqrt{x}} = \frac{5x^2 + 7}{2\sqrt{x}}$$

2. Let $f(x) = \frac{x+2}{x^2+4}$. Differentiate.

$$f'(x) = \frac{1(x^2+4)-2x(x+2)}{(x^2+4)^2} = \frac{x^2+4-2x^2-4x}{(x^2+4)^2)}$$

$$f'(x) = \frac{-x^2-4x+4}{(x^2+4)^2}$$

Implicit Differentiation

Implicit differentiation is a technique which facilitates the calculation of derivatives. It is an application of the chain rule.

Example

Find the slope of the tangent line to the circle $(x - 3)^2 + (y + 1)^2 = 37$ at the point (2,5).

Solution:

Differentiate on both sides keeping in mind that y depends on x

$$2(x - 3) + 2(y + 1)\frac{dy}{dx} = 0$$

$$\frac{dy}{dx} = \frac{3 - x}{y + 1}$$

For $x = 2$, $y = 5$ the slope is $\frac{3-2}{5+1} = \frac{1}{6}$

Practice Exercises

1. Let $f(x) = \frac{x^2}{x+1}$. Differentiate.

2. Let $f(x) = \sqrt{x}(x^2 + 1)^2$. Find $f'(x)$.

3. Let $f(x) = \frac{x}{\sqrt{x^2-1}}$. Find $f'(x)$.

4. Let $y = \left(x + \frac{1}{x}\right)^6$. Evaluate $\frac{dy}{dx}$.

5. Find the equation of the line tangent to the curve of $x^3 + y^3 = 9$ at $(1, 2)$.

6. Find the slope of the tangent line to the curve of $xy = y + 8$ at $(3, 4)$.

7. Let $1 + \frac{x}{y^2} = x - y$. Find $\frac{dy}{dx}$.

8. $3xy^3 = 4y - x$. Find $\frac{dy}{dx}$.

9. Find the slope of the tangent line to the curve of $(x + 1)^2 + (y - 4)^2 = 8$ at $(1, 6)$.

Part 3

3.1 Exponential and Logarithmic Functions

Properties of Exponents

For $a > 0$ and $b > 0$ and m, n rational numbers the following properties (laws) of exponents are valid:

1. $a^o = 1$

2. $a^m a^n = a^{m+n}$

3. $\frac{a^m}{a^n} = a^{m-n}$

4. $(a^m)^n = a^{mn}$

5. $a^m b^m = (ab)^m$

6. $1^m = 1$

7. $\left(\frac{a}{b}\right)^m = \frac{a^m}{b^m}$

8. $a^{-m} = \frac{1}{a^m}$

9. $a^{m/n} = \sqrt[n]{a^m}$

10. $\sqrt[n]{a^m} = \left(\sqrt[n]{a}\right)^m$

Definition of an Exponential Function

We call the function $y = a^x$ the exponential function with base a, $a > 0$ and $a \neq 1$, x is a real number.

The exponential function obeys the rules of the exponents.

The graph of the exponential function is <u>always rising</u>, <u>always positive</u>, (<u>never crosses the x-axis</u>) and the <u>y-intercept is 1</u> (since $a^o = 1$).

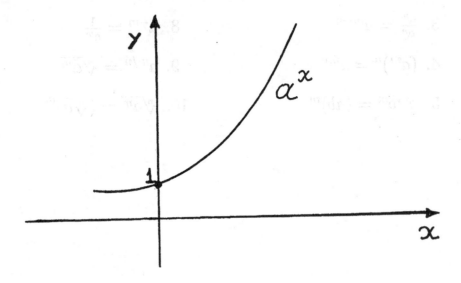

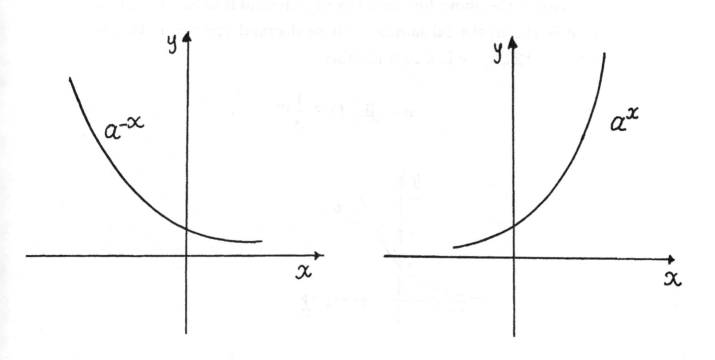

Negative expon. function
$y = a^{-x}$ always decreasing
domain $(-\infty, +\infty)$
$\lim\limits_{x \to \infty} a^{-x} = 0$
$\lim\limits_{x \to -\infty} a^{-x} = \infty$

positive expon. function
$y = a^{x}$ always increasing
domain $(-\infty, +\infty)$
$\lim\limits_{x \to \infty} a^{x} = \infty$
$\lim\limits_{x \to -\infty} a^{x} = 0$

The Natural Exponential Function

$$y = e^x$$

We call the above function the exponential function with base e (e is the irrational number whose decimal approximation is: $e \approx 2.7182\ldots$). x is a real number.

$$e = \lim_{n \to \infty} \left(1 + \frac{1}{n}\right)^n$$

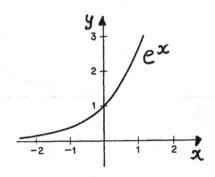

Logarithms

We say the logarithm of x to the base a is equal to y: ($\log_a x = y$) if and only if $a^y = x$. $a > 0$ and $a \neq 1$

$$\boxed{y = \log_a x \text{ means that } a^y = x}$$

We have seen that a a^y (expon. function) is always positive, therefore <u>only logarithms of positive numbers</u> can be found.

Properties of Logarithms

1. $\log(xy) = \log x + \log y$

2. $\log\left(\frac{x}{y}\right) = \log x - \log y$

3. $\log\left(x^y\right) = y \log x$

$$\log_{10} x = \log x \quad \cdot \quad \log_e x = \ln x$$

$$\ln x = \text{ natural logarithm of } x$$

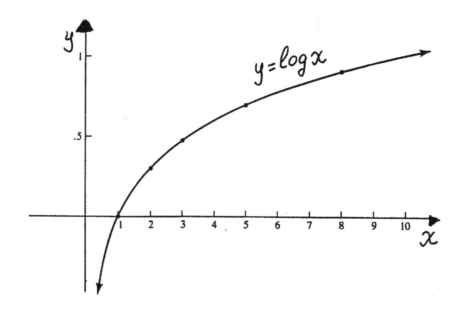

The <u>natural logarithmic function</u> $\boxed{\log_e x = \ln x}$

$$ln\ x\ = y \text{ if and only if } e^y = x$$

$$\boxed{ln\ x = y \text{ means that } e^y = x}$$

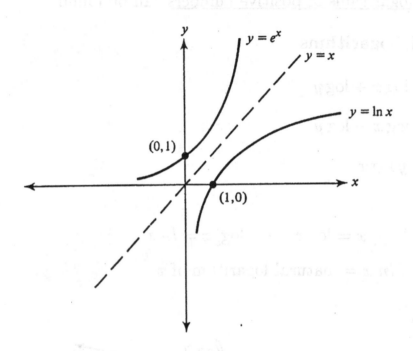

$$\boxed{ln\ 1 = 0} \qquad \boxed{ln\ e = 1}$$

Theorem 1

$$\log_a x = \frac{ln\ x}{ln\ a} \qquad a > 0, a \neq 1 \text{ for any positive } x \quad x > 0$$

Theorem 2

$$a^x = e^{x\ ln\ a}, \ a > 0 \text{ for any real} x$$

Derivatives of Exponential and Logarithmic Functions

If $u(x)$ is a differentiable function of x, then

$$\frac{d}{dx}(ln\ u) = \frac{1}{u}\frac{du}{dx} \text{ for } u > 0$$

$$\frac{d}{dx}(e^u) = e^u\frac{du}{dx}$$

Examples

1. Find the derivative of the function:

$$f(x) = ln\ (x^{10}),\ x > 0$$

Solution:

For $x > 0$

$$\frac{d}{dx}ln\ (x^{10}) = \frac{1}{x^{10}}\ \frac{d^{(x^{10})}}{dx} = \frac{10x^9}{x^{10}} = \frac{10}{x}$$

or

$$\frac{d}{dx}ln\ (x^{10}) = \frac{d}{dx}(10ln\ x) = \frac{10}{x}$$

2. Let $f(x) = ln\ \left(\frac{1-x}{x}\right)$. Find $\frac{df(x)}{dx}$.

Solution:

$$f(x) \;=\; ln\ (1-x) - ln\ x$$

$$\frac{df(x)}{dx} \;=\; \frac{-1}{1-x} - \frac{1}{x}.$$

3. Let $f(x) = \frac{ln\ x}{x}$. Find $f'(x)$.

Solution:

$$f'(x) = \frac{\frac{1}{x}\ x - ln\ x}{x^2} = \frac{1-ln\ x}{x^2}.$$

4. Let $f(x) = e^{x^2}$. Find $f'(x)$.

Solution:
$f'(x) = e^{x^2}(x^2)' = e^{x^2} \cdot 2x = 2xe^{x^2}$.
(e^{x^2} is a composite function, e^x is the outside function, x^2 is the inside function.)

5. Let $f(x) = ln\,(3x^2 + e^x)$. Differentiate

Solution:

$$\frac{df(x)}{dx} = \frac{1}{3x^2+e^x}(3x^2 + e^x)'$$

$$\frac{df(x)}{dx} = \frac{1}{3x^2+e^x}(6x + e^x)$$

$$\frac{df(x)}{dx} = \frac{6x+e^x}{3x^2+e^x}$$

6. Let $f(x) = xe^x$. Find all inflection points.

Solution:

$$f'(x) = e^x + xe^x = e^x(x+1)$$

$$f''(x) = e^x + e^x(x+1) = e^x(x+2)$$

The only candidate for an inflection points is $x = -2$.

$f(x)$ is concave down on $(-\infty, -2)$ and concave upon $(-2, \infty)$.

Inflection point: $\left(-2, -\frac{2}{e^2}\right)$.

Practice Exercises

1. Find the equation of the tangent line to the graph of $y = \ln (x + 1)$.

2. Let $f(x) = \ln \left[(x^2 + 1)(3x^2 - 2)\right]$. Find $f'(x)$. (**Hint:** Use $\ln (xy) = \ln x + \ln y$).

3. Let $f(x) = xe^x$. Use the first and second derivative to find any maximum, minimum and inflection point and sketch the graph of $f(x)$.

4. Find the equation of the tangent to $y = e^{-x}$ at $(0,1)$.

5. Let $f(x) = x^2 e^{-x}$. Use your graphic calculator to sketch $f(x)$. What can you say about the signs of $f'(x)$ and $f''(x)$.

6. Let $f(x) = e^{-\frac{x^2}{2}}$. Find the points of inflection. Sketch $f(x)$.

7. Let $f(x) = e^{x^2+2x}$. Differentiate $f(x)$.

8. Let $f(x) = x^x$. Differentiate $f(x)$. (**Hint:** $a^x = e^{x \ln a}$.)

9. Let $f(x) = \log_2(x^2 + 2x)$. Differentiate $f(x)$. (**Hint:** $\log_a x = \frac{\ln x}{\ln a}$)

10. Let $f(x) = \ln\left(\frac{x^4}{x-1}\right)$.

    On what intervals is $f(x)$ rising? falling?

    Find the extreme values of $f(x)$. Determine the concavity
    of the graph and find the points of inflection. Sketch the
    graph.

11. Let $f(x) = \ln(x^2 + 1)$. Find all local extreme points,
    inflection points. Sketch $f(x)$.

12. Let $f(x) = e^{-x^2}$. Find all relative extreme points. Find all inflection points. Sketch $f(x)$. (Use the first and second derivative).

13. Let $f(x) = e^x \ln x$. Differentiate $f(x)$.

14. Find the equation of the tangent line to the graph of $y = ln\,(x+1)$ at the point $(0,0)$.

15. Let $f(x) = \frac{ln\,x}{x}$, $x > 0$. Find all local maxima and minima, all inflection points (Use the first and second derivatives).

16. Let $f(x) = \ln \sqrt{\frac{1-x}{2-x}}$. Differentiate $f(x)$.

17. Let $f'(x) = (1-x)\,e^{-x}$. Use your graphing calculator to sketch $f'(x)$. What can you say about $f(x)$?

18. Use the first and second derivatives of $f(x) = \ln x$ to show that the graph is always rising and that it is everywhere concave down.

19. Let $f(x) = \ln (\ln x)$. Differentiate.

20. Let $f(x) = \ln \left(\frac{x^2+2}{x^4+7}\right)$. Differentiate.

3.2 Exponential Growth and Decay

Let $f(x) = C \cdot e^{Kx}$, where C and K are constants. Then $f'(x) = \frac{df(x)}{dx} = C \cdot e^{Kx} \cdot K = f(x) \cdot K$ or

$$\boxed{\frac{df(x)}{dx} = K \cdot f(x)}$$

We realize that the derivative (or the rate of change) of the function $f(x)$ is equal to the function multiplied by K, or we say:

The rate of change of the function is proportional to the function itself. This is a unique property that belongs to the exponential function only (because $(e^x)' = e^x$).

Therefore if a function $f(x)$ satisfies the relationship: $f'(x) = K \cdot f(x)$, then $f(x) = C \cdot e^{Kx}$

if $x = 0$, $f(0) = C \cdot e^0 = C$. Therefore C is the value of $f(x)$ at $x = 0$.

If we begin with the information that the rate of change of $f(x)$ is proportional to $f(x)$ itself, then $f(x)$ is an exponential function.

Examples

1. Continuously Compound Interest

   Let $P(t)$ be the principal in an account at time t. To say that interest is <u>compounded continuously</u> at rate r means that the time rate of growth of money is equal to the amount of money present multiplied by r. That is,

   $$P'(t) = \frac{dP(t)}{dt} = rP(t)$$

   This means:

   $$\boxed{P(t) = P_0 e^{rt}}$$

   Where P_0 is the initial investment. If the interest rate is $r = 10\%$ per year and one deposits \$1,000 initially, (a) How much money is present after 1 year and (b) How long does it take to double the original investment?

   Solution:

   (a)
   $$P(t) = P_0 e^{rt} = 1,000 e^{(.1)t}$$

   $$P(1) = 1,000 e^{.1} \approx 1,105$$

(b) We must find t such that

$$2000 = P(t) = 1000e^{(.1)t}$$

$$2 = e^{(.1)t}$$

$$\ln 2 = \ln e^{(.1)t} = (.1)t \ln e$$

$$\ln 2 = .1t$$

$$t = \frac{\ln 2}{.1} \approx 6.9 \text{ years}$$

2. The rate of growth of a bacteria colony is proportional to the size of the population. If the initial population is 100 and 2 hours later there are 150, how many are present at time 6 hours? It is given:

$$\frac{dp(t)}{dt} = P'(t) = K \cdot P(t)$$

this means:

$$P(t) \quad = \quad P_0 e^{Kt}$$

$$150 \quad = \quad P(2) = 100 e^{2K}$$

$$\frac{150}{100} \quad = \quad e^{2K}$$

$$\frac{3}{2} \quad = \quad e^{2k}$$

$$ln\left(\frac{3}{2}\right) \quad = \quad 2K$$

$$K \quad = \quad \frac{1}{2}ln\left(\frac{3}{2}\right)$$

$$P(t) \quad = \quad P_0 e^{Kt}$$

$$P(6) \quad = \quad 100 e^{\frac{6}{2}ln\left(\frac{3}{2}\right)} = 100 e^{ln\left(\frac{3}{2}\right)^3} = 100 \cdot \left(\frac{3}{2}\right)^3 = \frac{2700}{8}$$

3. The radioactive isotope iodine–131 is injected into the body to test how fast the thyroid gland absorbs iodine. It is known that I^{131} decays at a rate proportional to the amount present and that the proportionately factor is 8.6% per day. How long will it be before half of a given quantity has decayed?

Solution:

If $A(t)$ is the amount of I^{131} present at time t, then $\frac{dA(t)}{dt} = 0.086A(t)$. This means $A(t) = A_0 e^{-0.086t}$ where A_0 is the initial amount present. The questions is: for what t does $A(t) = \frac{1}{2}A_0$? We have

$$\frac{1}{2}A_0 \quad = \quad A_0 e^{-0.086t}$$

$$\frac{1}{2} \quad = \quad e^{-0.086t}$$

$$\ln\left(\tfrac{1}{2}\right) \quad = \quad \ln e^{-0.086t}$$

$$\ln 1 - \ln 2 \quad = \quad -0.086t$$

$$-\ln 2 \quad = \quad -0.086t$$

$$0.69 \quad = \quad 0.086t$$

$$t \quad \approx \quad 8$$

Therefore the half life of iodine–131 is about 8 days.

4. Find the amount of interest drawn by $100 compounded continuously at 6% for 5 years.

Solution:

$$P(t) = P_0 e^{rt} \qquad r = \text{annual interest rate}$$

$$P(5) = 100e^{(0.06).5} = 100e^{0.30} \approx 135$$

The interest is approximately $35.

Practice Problems

1. How long does it take a sum of money to double at 5% compounded continuously.

2. What is the present value of $1000 forty months from now? Assume continuous compounding at 6%.

3. In continuous compounding, what interest rate would be necessary in order that an initial $1000 investment grow to $1500 in 3 years?

4. Radioactive strontium–90 has a half-life of 27.9 years. An island has been deemed to have 5 times the maximum allowable level of strontium–90. How long will it be until the island is "safe" for habitation?

5. Suppose the amount $A(t)$ of a radioactive substance present at time t <u>decreases</u> at a route proportional to the amount present, that is

$$\frac{dA}{dt} = -KA(t), K > 0.$$

Half of the substance decays every 500 years. Suppose a sample of a painter's canvas is examined, and it is found that $\frac{1}{4}$ of the original substance has decayed. How old is the canvas?

Part 4

4.1 Anti-differentiation

The derivative of a function at $x = a$ gives the rate of change of the function at $x = a$, in other words <u>the derivative gives us information about the behavior of the function at that point.</u>

In the next several pages we are going to do the opposite, we are going to study the total effect or <u>cumulative effect of a function.</u> For example if we know the velocity of a moving object, we can find the total distance traveled by this object in a certain time interval.

<u>Anti means opposite.</u> Anti-differentiation is the opposite process of differentiation.

Let us consider a function $f(x) = x^3$, we are looking for another function $F(x)$ whose derivative is $f(x) = x^3$. Through trial and error we find that $F(x) = \frac{x^4}{4}$.

$$F'(x) = \left(\frac{x^4}{4}\right)' = \frac{1}{4}4x^3 = x^3.$$

The function $F(x)$ is called an anti-derivative of $f(x)$. Of course it is obvious that if we choose $F(x) = \frac{x^4}{4} + 5.31$ the derivative will be the same x^3.

$$F'(x) = \left(\frac{x^4}{4} + 5 \cdot 31\right)' = \frac{1}{4}4x^3 + 0 = x^3$$

Therefore the function $f(x) = x^3$ does not have only one anti-derivative but a whole set.

A function $F(x)$ is an anti-derivative of $f(x)$ if $F'(x) = f(x)$, that is if $f(x)$ is the derivative of $F(x)$.
The set of all the anti-derivatives of $f(x)$ is called the indefinite integral of $f(x)$ and the notation is:

$$\int f(x)dx = F(x) + C$$

$f(x)$ is called the integrand and dx is part of the notation and identifies the variable, C is the constant of integration.

Rules of Integration

1. $\int K\,dx = K\int dx = Kx + C$

2. $\int Kf(x)dx = K\int f(x)dx,\ K = \text{constant}$

3. $\int [f(x) \pm g(x)]\,dx = \int f(x)dx \pm \int g(x)dx$

4. Rule for power functions

$$\int x^r dx = \frac{x^{r+1}}{r+1} + C \quad \boxed{\text{if } r \neq -1}$$

$$\int x^{-1} dx = \int \frac{dx}{x} = \ln|x| + C \quad \boxed{\text{if } r = -1}$$

5. Rule for exponential function

$$\int e^{rx} dx = \frac{e^{rx}}{r} + C$$

Examples

1. Evaluate $\int \frac{7}{x} dx$.

Solution: $\int \frac{7}{x} dx = 7 \int x^{-1} dx = 7\ln|x| + C$
(x^{-1} is a power function, with $r = -1$)

2. Evaluate $\int (\sqrt{x} + 2x) dx$

Solution:

$$\int (\sqrt{x} + 2x) dx = \int \sqrt{x} dx + \int 2x dx =$$

$$= \int x^{\frac{1}{2}} dx + 2 \int x dx = \frac{x^{\frac{1}{2}+1}}{\frac{1}{2}+1} + 2\frac{x^2}{2} = \frac{x^{\frac{3}{2}}}{\frac{3}{2}} + x^2 =$$

$$= \frac{2}{3} x^{\frac{3}{2}} + x^2 + C$$

3. Assume $ln|2x+6| - e^2x$ is an anti-derivative of $f(x)$. Determine $f(x)$.

Solution:

It is given $F(x) = ln|2x+6| - e^2x$.

To find from the anti-derivative the function $f(x)$ we must act upon the anti-derivative with the opposite which is the derivative.

Therefore $F'(x) = f(x) = \left[ln|2x+6| - e^2x\right]' = \frac{2}{2x+6} - e^2$

4. The slope of a curve $y = f(x)$ at each point x is $4x^3 + x^{-2}$. What are all possible functions with this property?

Solution: It is given $m(x) = 4x^3 + x^{-2} = f'(x)$

$$f(x) = \int f'(x) = \int (4x^3 + x^{-2})dx = x^4 - \frac{1}{x} + C$$

5. Determine $f(x)$ where:

$$\int f(x)dx = \frac{d}{dx}\left(e^{2x} - 4x^3\right) + C'$$

Solution:

To cancel the effect of the integral on $f(x)$ we must act upon the integral with the opposite which is the derivative. Therefore

$$\frac{d}{dx}\int f(x)dx = \frac{d}{dx}\left[\frac{d}{dx}\left(e^{2x} - 4x^3\right) + C\right]$$

$$f(x) = \frac{d}{dx}\left(2e^{2x} - 12x^2 + C\right)$$

$$f(x) = 4e^{2x} - 24x$$

6. Evaluate $\int \frac{x^4+x^2+8}{4x^3}dx$

Solution:

$$\int \frac{x^4+x^2+8}{4x^3}dx = \int \frac{x^4}{4x^3}dx + \int \frac{x^2}{4x^3}dx + \int \frac{8}{4x^3}dx =$$

$$= \int \frac{x}{4}dx + \int \frac{x^{-1}}{4}dx + \int 2x^{-3}dx = \frac{x^2}{8} + \frac{1}{4}ln\,|x| - x^{-2} + C$$

7. Suppose that the marginal cost for producing x items of a certain product is given by:

$$2000 + 10x^{3/2} \text{ dollars per item}$$

If the fixed costs are \$4,000, what is the cost of producing x items.

$$C(x) = \int (2000 + 10x^{3/2})dx = \int 2000dx + \int 10x^{3/2}dx =$$

$$= 2000x + 10\frac{x^{5/2}}{\frac{5}{2}} = 2000x + 4x^{5/2} + C$$

$$C(x) = 2000x + 4x^{5/2} + C$$

$$C(0) = 2000(0) + 4(0)^{5/2} + C = 4000$$

$$C = 4000$$

The cost is for producing x items

$$C(x) = 2000x + 4x^{5/2} + 4000$$

8. Suppose the velocity of an object is $v(t) = 6t^2 - 8t$, with a position -5 when time 0. Find $s(t)$.

We know that $v(t) = \frac{ds(t)}{dt}$ or

$$ds(t) = v(t)dt$$

$$s(t) = \int ds(t) = \int v(t)dt$$

Therefore $s(t) = \int (6t^2 - 8t)dt$

$$s(t) = \frac{6t^3}{3} - \frac{8t^2}{2} + C$$

$$s(t) = 2t^3 - 4t^2 + C$$

$$s(0) = 0 - 0 + C = -5$$

Therefore $s(t) = 2t^3 - 4t^2 - 5$

Practice Exercises

1. Evaluate $\int \frac{x^2+1}{x}dx$ $= \int \frac{x^2}{x}dx + \int \frac{1}{x}dx$

$$\int x^{-1}dx = \frac{x^{-1+1}}{-1+1} + C$$

2. Find $\int (e^{3x} + 5x^{-4/3})dx$

3. An object is dropped from the top of the 1100-foot tall Sears Tower in Chicago with initial velocity–20 ft /sec. and constant acceleration $a(t) = -32$ ft /sec^2. Find the distance of the object from the top of the building after t seconds.

4. Suppose a company has found that the marginal cost at a level of production x is given by $C'(x) = \frac{20}{\sqrt{x}}$ and the fixed cost is \$20,000. Find the cost function $C(x)$.

5. The tangent line to the graph of a function $f(x)$ at $(x, f(x))$ has slope $\frac{3}{x^2}$. Given that $(-1, 5)$ lies on the graph, determine $f(x)$.

6. Determine $f(x)$ if $f''(x) = x^2 - 10x + 1$, $f'(3) = -23$ and $f(0) = -2$.

7. The current selling price of a piece of property is $40,000. Its value will appreciate at a rate of $3000 + 180\sqrt{x}$ dollars per year, x years hence.

   (a) How much does its value increase in 4 years?

   (b) How much does its value increase in 1 year? Why is this less than the growth rate of its value at the end of the first year?

8. For a particular urban group, sociologist studied the average yearly income y (in dollars) that a person can expect to receive with x years of education before seeing regular employment. They estimated that the rate at which income changes with respect to education is given by

$$\frac{dy}{dx} = 10x^{3/2} \qquad 4 \leq x \leq 16$$

   where $y = 5872$ when $x = 9$. Find y.

9. The marginal cost function of a production process is given by $C'(x) = \frac{1}{60}x^2 - x + 615$. The fixed costs of production are $1000.

   (a) Find the cost-output function.

   (b) Find the total cost of producing 30 units.

4.2 Areas and the Definite Integral

Let $f(x) = x^2$. We can approximate the area under the graph of the function $f(x)$ over the x-axis on the interval [0,2] by evaluating the fourth Reimann Sum S_4

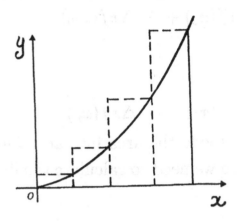

$$\boxed{\Delta x = \frac{b-a}{n}}$$

$$\Delta x = \frac{2-0}{4} = \frac{1}{2}$$

we divide the interval from $x = 0$ to $x = 2$ into four equal length subintervals and then create four rectangles. Then we find the area of each rectangle and add up the four areas.

$$S_4 = \text{area of rectangle 1} + \text{area of rectangle 2} +$$
$$\text{area of rectangle 3} + \text{area of rectangle 4}$$

$$S_4 = \Delta x f(x_1) + \Delta x f(x_2) + \Delta x f(x_3) + \Delta x f(x_4)$$

$$S_4 = \Delta x \left[f(x_1) + f(x_2) + f(x_3) + f(x_4) \right]$$

$$S_4 = \frac{1}{2} \left(\frac{1}{4} + 1 + \frac{9}{4} + 4 \right) = \frac{1}{2} \left(\frac{30}{4} \right) = \frac{30}{8}$$

The fourth Reimann sum is an approximation to the area, $A \approx S_4$. Obviously this is a poor approximation. If we want a good approximation we must divide the area into a greater number of rectangles, for example

$$S_{20} = \Delta x f(x_1) + \Delta x f(x_2) + \cdots \Delta x f(x_{20})$$

or

$$S_n = \Delta x f(x_1) + \Delta x f(x_2) + \cdots \Delta x f(x_n)$$

The exact area will result if we divide the area into an infinite number of rectangles, which mean we need to calculate the limit at infinity for S_n.

$$A = \lim_{n \to \infty} S_n = \lim_{n \to \infty} [\Delta x f(x_1) + \Delta x f(x_2) \cdots + \Delta x f(x_n)]$$

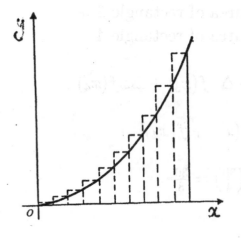 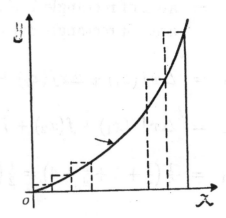

Definition of the Area Under the Graph of a Function

Let $f(x)$ be a non-negative continuous function on $[a, b]$, then the limit at infinity of the nth Reimann Sum (if it exists) is equal to the exact area under the graph of $f(x)$ over the interval $[a, b]$.

$$A = \lim_{n \to \infty} [\Delta x f(x_1) + \Delta x f(x_2) + \cdots \Delta x f(x_n)]$$

Where $x_1, x_2, \cdots x_n$ are points in the n subintervals of $[a, b]$. This limit is defined as the definite integral of $f(x)$ from a to b and the notation is

$$\int_a^b f(x)dx.$$

Therefore:

$$\int_a^b f(x)dx = \lim_{n \to \infty} [\Delta x f(x_1) + \Delta x f(x_2) + \cdots \Delta x f(x_n)] = A$$

The definite integral $\int_a^b f(x)dx$ is a <u>number</u>.

The beginning of the interval a is called the lower limit of integration and the end of the interval b is called the upper limit of integration.

The Fundamental Theorem of Calculus

Let $f(x)$ be a continuous function on the closed interval $[a, b]$ and $F(x)$ be an anti-derivative of $f(x)$ on $[a, b]$. Then

$$\int_a^b f(x)\,dx = F(b) - F(a)$$

$F(b) - F(a)$ is denoted by $F(x) \, \big]_a^b$.

Examples

1. Evaluate the definite integral $\int_1^5 3x^2 dx$.

   <u>Solution:</u> $\int_1^5 3x^2 dx = \frac{3x^3}{3} \,]_1^5 = x^3 \,]_1^5 = 5^3 - 1^3 = 124$

2. Evaluate $\int_0^1 e^x dx$

   <u>Solution:</u> $\int_0^1 e^x dx = e^x \,]_0^1 = e' - e^0 = e - 1$

3. Evaluate $\int_1^4 (2x^3 - 1)dx$

   <u>Solution:</u> $\int_1^4 (2x^3-1)dx = \left(\frac{2x^4}{4} - x\right) \,]_1^4 = \left(\frac{4^4}{2} - 4\right) - \left(\frac{1^4}{2} - 1\right) = \frac{249}{2}$

Practice Exercises

1. Evaluate $\int_2^7 \frac{1}{r^2} dr$

2. Evaluate $\int_1^4 \frac{2}{\sqrt{x}} dx$

3. Evaluate $\int_1^2 \frac{x^2 - 2x + 5}{x} dx$

4. A water storage tank has a leak and is losing water at the rate of 2t-1000 gallons per hour, where t is the number of hours since the tank began to leak. How much water will be lost between $t = 10$ and $t = 20$?

5. The production rate of a company on the tth day after it opens is given by $\frac{3t}{10}+10$ items per day. How many items will the company produce between the twentieth and thirtieth day after it opens?

6. Find the change P in a population between times t_1 and t_2 if $G(t)$ is its growth rate at time t.

7. If after t days a rumor is spreading at the rate of $100e^{-2t}$ new people per day, how many people hear of the rumor during the fifth and sixth days; that is, from $t = 4$ to $t = 6$?

8. If, t hours into a 24 hour work day, $(25 + 2t)$ tons of waste per hour are being dumped into a river, how many tons are dumped during the whole day?

Area by Integration

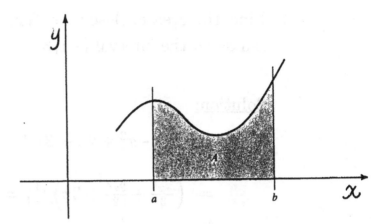

The area of the region between the continuous non-negative function $f(x)$ and the x-axis bounded by $x = a$ and $x = b$ is given by

$$A = \int_a^b f(x)dx$$

Examples

1. Find the area under the graph of $f(x) = x^2$ over the x-axis on the interval $[0,2]$.

   Solution: $\int_0^2 x^2 dx = \frac{x^3}{3}\big]_0^2 = \frac{2^3}{3} - \frac{0^3}{3} = \frac{8}{3}$.

   This number ($\frac{8}{3}$ in square units) is the exact area under the graph of x^2 over the x-axis on the interval $[0,2]$. Compare this number with $\frac{30}{8}$ which was an approximation to the same area by calculating the 4th Reimann sum.

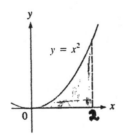

2. Find the area enclosed by $f(x) = -x^2 + 2x + 3$ and the x-axis on the interval $[-1, 3]$.

Solution:

$$A = \int_{-1}^{3} \left(-x^2 + 2x + 3 \right) dx =$$

$$= \left(\frac{-x^3}{3} + \frac{2x^2}{2} + 3x \right) \Big]_{-1}^{3} =$$

$$= \left(\frac{-27}{3} + 9 + 9 \right) - \left(\frac{1}{3} + 1 - 3 \right) = \frac{32}{3} \text{ square units}$$

$$F(3) - F(-1)$$

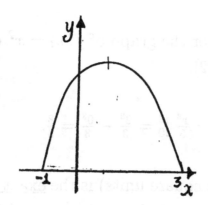

The Area of a Region Between Two Curves

Let $f(x)$ and $g(x)$ be continuous functions on the interval $[a, b]$ and $f(x) \geq g(x)$ for each x in this interval. The area of the region bounded by the graphs of $f(x)$ and $g(x)$ and the vertical lines $x = a$ and $x = b$ is given by:

$$\int_a^b [f(x) - g(x)]\, dx.$$

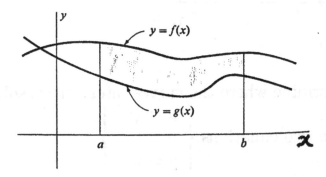

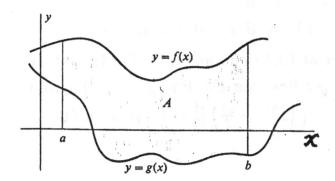

Examples

1. Find the area of the region bounded by the curves $y = \sqrt{x}$ and $y = x$.

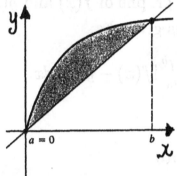

Solution: To determine where the curves intersect we solve the system of the two equations $\left\{ \begin{array}{c} y = \sqrt{x} \\ y = x \end{array} \right\}$

$$
\begin{aligned}
\sqrt{x} &= x \\
x &= x^2 \\
x - x^2 &= 0 \\
x(1 - x) &= 0, x = 0, x = 1.
\end{aligned}
$$

The curves intersect at $(0, 0)$ and at $(1, 1)$. On the interval $[0, 1]$ the parabola $\sqrt{x}$ lies over the line $y = x$. Therefore $A = \int_0^1 (\sqrt{x} - x)dx = \left(\frac{2}{3}x^{3/2} - \frac{x^2}{2} \right)]_0^1 = \frac{1}{6}$ square units.

2. Find the area between the curves $y = x^2$ and $y = -x + 6$.

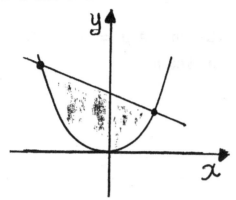

Solution. We first find the limits of integration by finding the points of interaction.

$$\left\{ \begin{array}{ll} y & = & x^2 \\ y & = & -x + 6 \end{array} \right\}$$

$$x^2 = -x + 6$$

$$x^2 + x - 6 = 0$$

$$(x + 3)(x - 2) = 0, \quad x = -3, 2$$

Therefore

$$A = \int_{-3}^{2} \left(-x + 6 - x^2 \right) dx = \left(\frac{-x^2}{2} + 6x - \frac{x^3}{3} \right) \bigg]_{-3}^{2}$$

$$= \frac{22}{3} - \left(-\frac{27}{2} \right) = \frac{125}{6} \text{ square units}$$

Practice Exercises

1. Find the area between the curves $y = x^2 - 2x + 2$ and $y = -x^2 + 6x - 4$. Make a sketch.

2. Find the area between the curves $y = x + 5$ and $y = x^2 - 1$ on the interval $[-1, 2]$. Make a sketch.

3. Find the area bounded by the curves $y = 4x - x^2 + 8$ and $y = x^2 - 2x$. Make a sketch.

4. Find the area of the region from $x = 1$ to $x = 2$ bounded by the graphs of $f(x) = x + 2$ and $g(x) = 5 - 2x$. Make a sketch.

5. Find the area of the region bounded by the graph of $f(x) = 4 - x^2$, the axis and the vertical line $x = -1$ and $x = 3$. Sketch the region.

6. Sketch the region enclosed by the graphs of $y = 2x^2 - 4x - 7$ and $y = 9$ and determine its area.

4.3   Techniques of Integration

A. Integration by Substitution

This technique is used for integrals of the form:

$$\int f(g(x)) \cdot g'(x) dx$$

$f(g(x))$ is a composite function

$g'(x)$ is the derivative of the inside function.

The technique involves simplifying the integral by changing the variable.

Let $u = g(x)$ then $\frac{du}{dx} = g'(x)$ and $du = g'(x)dx$

Therefore:

$$\int f(g(x))g'(x) dx = \int f(u) \cdot du$$

The integral: $\int f(u) du$ is much simpler than the original and can be evaluated now by using the integration rules.

Examples

1. Evaluate: $\int (4x + 4) \cdot e^{x^2 + 2x} dx = \int 2(2x + 2) \cdot e^{x^2 + 2x} dx$.

   Solution:Let

   $$u \quad = \quad x^2 + 2x$$

   $$\frac{du}{dx} \quad = \quad 2x + 2$$

   $$du \quad = \quad (2x + 2)dx$$

   Therefore:

   $$2\int (2x + 2)e^{x^2 + 2x} dx = 2\int e^u du = 2e^u = 2e^{x^2 + 2x} + K$$

Practice Exercises

1. Evaluate: $\int 12x^3(x^4+1)^5 dx$

2. Evaluate: $\int \frac{2x}{(x^2-3)^6} dx$

3. Evaluate: $\int \frac{\ln x}{x} dx$

4. Evaluate: $\int \frac{e^{\sqrt{x}}}{x} dx$

5. Evaluate: $\int \frac{e^x - e^{-x}}{e^x + e^{-x}} dx$

6. Evaluate: $\int \frac{3x^2 + x}{4x^3 + 2x^2 + 5} dx$

B. Integration by Parts

Let us start with the product rule:

$$[f(x) \cdot g(x)]' = f'(x) \cdot g(x) + g'(x)f(x)$$

Now let us <u>undo</u> the product rule, which means we would like to cancel the differentiation on the product $f(x) \cdot g(x)$. <u>This can be done with the opposite process of differentiation which is integration (or anti-differentiation).</u>

$$\int [f(x) \cdot g(x)]' = \int f'(x)g(x)dx + \int g'(x)f(x)dx$$

$$f(x) \cdot g(x) = \int f'(x) \cdot g(x)dx + \int g'(x)f(x)dx$$

Let us treat the integral $\int f(x)g'(x)dx$ as an unknown, then

$$\int f(x)g'(x)dx = f(x)g(x) - \int f'(x)g(x)dx + K$$

This formula is called the <u>integration by parts formula</u>.

Example

Evaluate: $\int xe^{2x}dx$

<u>Solution:</u>

Let $f(x) = x$ and $g'(x) = e^{2x}$. Then $f'(x) = 1$ and $g(x) = \int e^{2x}dx = \frac{e^{2x}}{2}$.

Substitution in the integration by parts formula yields

$$\int xe^{2x}dx = x\frac{e^{2x}}{2} - \int 1 \cdot \frac{e^{2x}}{2}dx =$$

$$= x\frac{e^{2x}}{2} - \frac{e^{2x}}{4} + K$$

Practice Exercises

1. Evaluate: $\int (6x+1)e^{3x}dx$

2. Evaluate: $\int x\ln x\,dx$

3. Evaluate: $\int \frac{\ln x}{\sqrt{x}}dx$

4. Evaluate: $\int x^2 e^{-x} dx$

5. Evaluate: $\int 2x^3 (ln\ x)^2 dx$

1. Given that $g(x+h) - g(x) = \frac{1}{x+h} - \frac{1}{x}$, find $g'(3)$.

   (a) x^2 (c) does not exist   (e) $\frac{1}{4}$

   (b) $-\frac{1}{3}$ (d) $-\frac{1}{9}$

2. Let

$$f(x) = \begin{cases} x^2 + 1, & \text{if } x \leq 0 \\ -4 & \text{if } x = 0 \\ 5x + 1 & \text{if } x > 0 \end{cases}$$

   (a) $f(x)$ is continuous at $x = 0$

   (b) $f(x)$ is continuous but non differentiable at $x = 1$

   (c) $f(x)$ has no limit as x approaches 0

   (d) $f(x)$ is not defined at $x = 0$.

   (e) $f(x)$ is discontinuous at $x = 0$.

3. Let $f(x) = 1 + x^{\frac{2}{3}}$

   At $x = 2$

   (a) $f(x)$ is not defined

   (b) $f(x)$ has a local maximum

   (c) $f(x)$ has a local minimum

   (d) $f'(x)$ is negative

   (e) $f(x)$ is increasing

4. Let $6x^2 + 6x - 12$ be the derivative of a function $f(x)$. At $x = -4$

   (a) $f(x)$ is increasing

   (b) $f'(x)$ is negative

   (c) $f(x)$ is decreasing

   (d) $f'(x)$ is equal to 2000

   (e) $f(x)$ has a local minimum

5. In the figure below find the equation of the tangent line to $f(x)$ at the point A.

(a) $y = \dfrac{1}{2}x + 8$

(b) $y = \dfrac{x}{2} + 3$

(c) $y = x + 4$

(d) $y = 1$

(e) None of the above

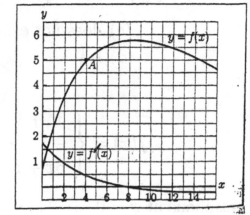

6. Suppose the revenue from selling x custom-made office desks is: $R(x) = 2000(1 - \frac{1}{x+1})$ dollars. Find the marginal revenue when x desks are sold.

(a) $\dfrac{2000}{(x+1)^2}$

(b) $-2000(x+1)^{-1}$

(c) ∞

(d) Does not exist

(e) $2000 - \dfrac{2000}{(x+1)^2}$

7. A model rocket is launched vertically such that its distance s (in feet) from the ground at any time t (in seconds) is $s(t) = -16t^2 + 480t$. Which of the following statements is true?

(a) The rocket is always rising

(b) The rocket is always falling

(c) The rocket is falling when $t = 0$

(d) The rocket neither rises nor falls

(e) The rocket is rising $t < 15$

8. Suppose a car is travelling on a straight road and $s(t)$ is the distance travelled after t hours. If the car is moving forward but slowing down at time $t = a$ how could you describe this information?

(a) $s'(t)$ is positive and $s''(t)$is negative at $t = a$

(b) $s'(t)$ is negative at $t = a$

(c) $s''(t)$ is positive at $t = a$

(d) $s'(t)$ is a positive constant function

(e) $s(t)$ is a constant function

140

9. The following figure contains the graph of $f'(x)$, derivative of the function $f(x)$. At $x = 4$, $f(x)$.

(a) is increasing

(b) has a relative minimum

(c) has a relative maximum

(d) is decreasing

(e) is undefined

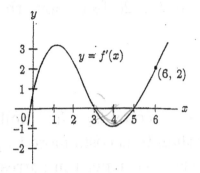

Let $y = \frac{x+1}{x+2}$.

10. Describe the behavior of $f(x)$ as x approaches -2 from the left:

(a) $f(x)$ increases with no bound   (e) None of the above.

(b) $f(x)$ decreases with no bound

(c) $f(x)$ has a limit which is 2

(d) $f(x)$ has a limit which is 0

11. Find $\lim\limits_{x \to 1} \dfrac{\sqrt{x} - 1}{x - 1}$

(a) $\frac{1}{2}$

(b) $\frac{1}{4}$

(c) does not exist

(d) ∞

(e) 2

12. As h approaches 0 what value is approached by $\frac{(3+h)^2 - (3)^2}{h}$?

(a) 0

(b) No value

(c) 6

(d) $3 + h$

(e) h

141

13. A helicopter is rising straight up in the air. Its distance from the ground t seconds after takeoff is $s(t)$ feet, where $s(t) = t^2 + t$. Find the velocity of the helicopter when it is 20 feet above the ground.

(a) 0 ft/sec

(b) 9 ft/sec

(c) 11 ft/sec

(d) -32 ft/sec ^{2}

14. Suppose it costs a company $C(x)$ dollars to produce x tons of steel in a mill. For h greater than 0, it costs more to produce $x + h$ tons in a week. Then the average increase in cost per ton corresponding to this change is given by:

(a) $C'(x)$

(b) $\lim\limits_{h \to 0} \dfrac{C(x+h) - C(x)}{h}$

(c) The marginal cost

(d) $\dfrac{C(x+h) - C(x)}{h}$

(e) $\dfrac{dC(x)}{dx}$

15. It is estimated that x months from now, the population of a certain community will be

$$P(x) = x^2 + 20x + 8000.$$

At what rate will the population be changing with respect to time 15 months from now?

(a) 20 people per month

(b) -20 people per month

(c) 50 people per month

(d) 8,050 people per month

(e) 30 people per month

ANSWERS

EXAM 1

1 (A) (B) (C) ● (E)   11 ● (B) (C) (D) (E)
2 (A) (B) (C) (D) ●   12 (A) (B) ● (D) (E)
3 (A) (B) (C) (D) ●   13 (A) ● (C) (D) (E)
4 ● (B) (C) (D) (E)   14 (A) (B) (C) ● (E)
5 (A) ● (C) (D) (E)   15 (A) (B) ● (D) (E)
6 ● (B) (C) (D) (E)   16 (A) (B) (C) (D) (E)
7 (A) (B) (C) (D) ●   17 (A) (B) (C) (D) (E)
8 ● (B) (C) (D) (E)   18 (A) (B) (C) (D) (E)
9 (A) (B) (C) ● (E)   19 (A) (B) (C) (D) (E)
10 ● (B) (C) (D) (E)   20 (A) (B) (C) (D) (E)

1. Using the graph below find the following limit

$$\lim_{x \to \infty} (2f(x) - 1)$$

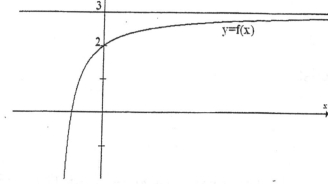

(a) 6

(b) 5

(c) 2

(d) 0

(e) ∞

2. If $f(t) = \dfrac{1}{t^{\frac{2}{3}}}$ then $\displaystyle\lim_{h \to 0} \dfrac{f(-8+h) - f(-8)}{h}$ equals

(a) $f'(8)$

(b) $\dfrac{1}{48}$

(c) $f'(0)$

(d) It does not exist.

(e) $-\dfrac{2}{96}$

3. Determine the slope of the tangent line to the graph of the function $f(x) = \sqrt{x^2 + 2x + 9}$ at the point $x = 0$.

(a) 1

(b) $\frac{1}{3}$

(c) 3

(d) $\sqrt{2}$

(e) It does not exist.

4. Let $f(x) = \begin{cases} 2x + 2 & \text{if } x < -1 \\ 2 - x^2 & \text{if } -1 \leq x \leq 1 \\ 4 - 2x & \text{if } x > 1 \end{cases}$

Evaluate $f'(1)$

(a) 0

(b) 2

(c) does not exist

(d) 4

(e) -2

5. The tangent line to the graph of $y = \dfrac{1}{x}$ at the point $P\left(a, \frac{1}{a}\right)$ is perpendicular to the line $y = 4x + 1$.

Find the coordinates of a point P in the first quadrant.

(a) $\left(\frac{1}{4}, 4\right)$

(b) $\left(2, \frac{1}{2}\right)$

(c) $\left(-\frac{1}{2}, -2\right)$

(d) $(-2, -7)$

(e) $\left(4, -\frac{1}{16}\right)$

6. An equation for the line tangent to the graph of $y = \sqrt{x}$ at the point on the graph with $x = 4$ is:

(a) $y = 2x - 6$

(b) $y = \frac{1}{4}x$

(c) $y = \frac{1}{4}x + 1$

(d) $y = x - 2$

(e) $y = 0$

7. Suppose $g(3) = 2$, $g'(3) = 4$. Find $f'(3)$, where $f(x) = 3[g(x)]^3$.

(a) 36

(d) 108

(b) 96

(e) none of the above

(c) 144

8. For a function $f(x)$, the following relation is given:

$$\lim_{x \to a} f(x) = f(a) \text{ where } a = \text{ real number}$$

Then, which of the following is NOT ALWAYS true?

(a) $\lim_{x \to a} f(x)$ exists.

(b) $f(x)$ is defined at $x = a$.

(c) The function $f(x)$ is continuous at $x = a$.

(d) $\lim_{x \to a^-} f(x) = \lim_{x \to a^+} f(x)$.

(e) $f'(a)$ exists.

9. Let $R(x)$ be the revenue generated from producing and selling x units of goods. What would you do to find the marginal revenue from producing 1000 units of goods?

(a) Compute $R(1000)$.

(b) Set $R(x) = 1000$ and solve for x.

(c) Compute $R'(1000)$.

(d) Find a value of x for which $R'(x) = 1,000$.

(e) Compute $\dfrac{R(x + h) - R(x)}{h}$

10. A rock is thrown off a cliff. Its distance from the ground below at t seconds is $s(t) = -16t^2 - 16t + 96$ feet. How fast will the rock be falling when it hits the ground?

(a) $-32\ ft/sec^2$

(b) $-16\ ft/sec$

(c) $-80\ ft/sec$

(d) $0\ ft/sec$

(e) $-32t$

11. The figure below shows the graph of $y = f'(x)$, the derivative of a function $f(x)$. For what value of x does the graph of the original function $f(x)$ have a horizontal tangent line?

(a) $x = 0$

(b) $x = 1$

(c) $x = 2$

(d) $x = 3$

(e) $x = 4$

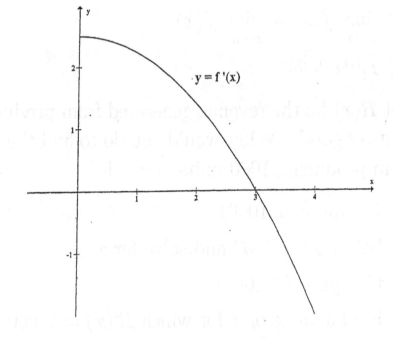

12. In the figure below the straight line is tangent to the graph of $f(x)$. Find $f'(4)$.

(a) $\frac{1}{2}$

(b) does not exist

(c) -8

(d) -4

(e) 3

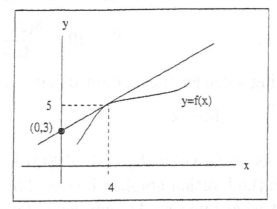

13. The function defined by $f(x) = \begin{cases} x^2 + 1, & x < 1 \\ x^3 + 4x - 3, & x \geq 1 \end{cases}$

(a) has no limit at $x = 4$

(b) has no limit at $x = 1$

(c) is not continuous at $x = 1$

(d) is continuous at $x = 1$

(e) is differentiable at $x = 1$

14

The amount of salt, S, in an aquarium tank at time t is given by the function

$$S = 10 - \frac{6(2t+1)}{4t+1}.$$

The limiting value for the amount of salt as $t \to \infty$ is:

(a) 3   (b) $-\infty$ (c) 7   (d) Undefined   (e) 10

15. The number of farms in the United States t years after 1925 is $f(t)$ million where f is the function graphed below. The graphs of $f'(t)$ and $f''(t)$ are also shown. At what rate was the number of farms declinning in 1990?

(a) 2 million farms per year

(b) the rate was constant

(c) the rate was zero

(d) 30,000 farms per year

(e) 50,000 farms per year

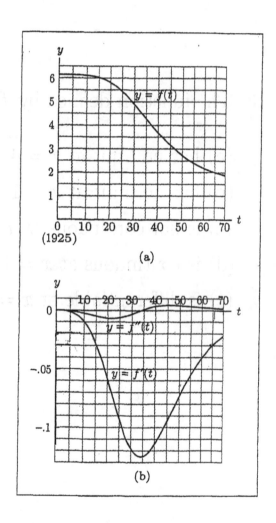

(a)

(b)

150

ANSWERS

EXAM 2

1. (A) ● (C) (D) (E)   11. (A) (B) (C) ● (E)
2. (A) ● (C) (D) (E)   12. ● (B) (C) (D) (E)
3. (A) ● (C) (D) (E)   13. (A) (B) (C) ● (E)
4. (A) (B) ● (D) (E)   14. (A) (B) ● (D) (E)
5. (A) ● (C) (D) (E)   15. (A) (B) (C) ● (E)
6. (A) (B) ● (D) (E)   16. (A) (B) (C) (D) (E)
7. (A) (B) ● (D) (E)   17. (A) (B) (C) (D) (E)
8. (A) (B) (C) (D) ●   18. (A) (B) (C) (D) (E)
9. (A) (B) ● (D) (E)   19. (A) (B) (C) (D) (E)
10. (A) (B) ● (D) (E)   20. (A) (B) (C) (D) (E)

1. Given that $g(x + h) - g(x) = \dfrac{1}{(x+h)^2} - \dfrac{1}{x^2}$ find $g'(-3)$.

   (a) $2x^3$

   (b) $\dfrac{2}{27}$

   (c) 1

   (d) $\dfrac{1}{3}$

   (e) $-\dfrac{2}{9}$

2. Suppose that the weight in milligrams of a cancerous tumor at the time t is $W(t) = (0.1)t^2$, where t is measured in weeks. At what time is the tumor growing at a rate of 5 milligrams per week?

   (a) 1 week
   (b) never
   (c) 25 weeks
   (d) 2.5 weeks
   (e) 2 weeks

3. Decide whether the following statements are true or false for any function $f(x)$.

    I). If $f(x)$ is defined at $x = a$, then $f(x)$ is continuous at $x = a$.

    II). If $f(x)$ is continuous at $x = a$, then $f(x)$ is defined at $x = a$.

    III). If $f(x)$ is continuous at $x = a$, then the limit of $f(x)$ as x approaches a exists.

    IV). If the limit of $f(x)$ as x approaches a exists, then $f(x)$ is continuous at $x = a$.

(a) II) and III) are always true.

(b) II) and IV) are always true.

(c) I), II), III), IV) are always false.

(d) I), II), III), IV) are always true.

(e) none of the above

4. Let

$$f(x) = \begin{cases} x + 1 & \text{for } x < 0 \\ (x + 1)^2 & \text{for } 0 \le x < 2 \\ 2 - x & \text{for } x \ge 2 \end{cases}$$

At $x = 0$, the function $f(x)$ is:

(a) Continuous and differentiable.

(b) Discontinuous and non-differentiable.

(c) Continuous and non-differentiable.

(d) Discontinuous and differentiable.

(e) None of the above.

5. Find the equation of the tangent line to the curve

$$y = \frac{8}{x^2 + x + 2} \text{ at } x = 2$$

(a) $y = -\frac{5}{8}x + \frac{9}{4}$

(b) $y = \frac{5}{8}x - \frac{9}{8}$

(c) $y = -5x + 11$

(d) $y = 5x + \frac{9}{4}$

(e) None of the above.

6. The $\lim_{h \to 0} \dfrac{\sqrt{8 - x - h} - \sqrt{8 - x}}{h}$ is equal to:

(a) $\sqrt{x - 8}$

(b) $\frac{1}{2}8^{-\frac{1}{2}}$

(c) $\dfrac{-1}{2\sqrt{8 - x}}$

(d) $\dfrac{1}{\sqrt{x - 8}}$

(e) $\dfrac{d}{dx}(-\sqrt{8 - h})$

7. The total cost (in thousands of dollars) of producing x units of a certain commodity is $C(x) = 6x^2 + 2x + 10$. Estimate the additional cost incurred if the production level is increased from 10 to 10.5 units.

(a) $128,000$

(b) $693,000$

(c) $120,000$

(d) $61,000$

(e) $6,000$

8. Let $f(x) = \sqrt{2x+1}$. Then $\lim\limits_{h \to 0} \dfrac{\sqrt{8+2h+1}-3}{h}$ equals

(a) does not exist

(b) $f'(3)$

(c) $f'(9)$

(d) $\dfrac{1}{3}$

(e) $f'(8)$

9. The tangent line to the curve $f(x) = \dfrac{1}{3}x^3 - 4x^2 + 18x + 22$ is parallel to the line $6x - 2y = 1$ at the point(s)

(a) $x = 2$

(b) $x = 3,\ x = 5$

(c) $x = 5$

(d) $x = -3,\ x = 5$

(e) $x = 2,\ x = 5$

10. Let $f(x) = \dfrac{2x}{x^2 + 3x}$

Determine the horizontal and the vertical asymptotes to the graph of $f(x)$.

(a) $x = 0,\ y = 2$ (d) $x = 2,\ y = -3$

(b) $x = -3,\ y = 0$ (e) $x = 0,\ y = \dfrac{2}{3}$

(c) $x = 2,\ y = 0$

11. Determine the value of c that makes the function $f(x)$ continuous on the entire set of real numbers.

$$f(x) = \begin{cases} x - 2, & x \le 5 \\ c \cdot x - 3, & x > 5 \end{cases}$$

(a) 0

(b) $\dfrac{6}{5}$

(c) 1

(d) $\dfrac{5}{6}$

(e) 6

12. The line $y = 6x - b$ is tangent to the graph of the function $f(x) = \frac{1}{2}x^3$ at the point x, where $x > 0$. Find the value of b.

(a) 4

(b) $\dfrac{1}{2}$

(c) 1

(d) 8

(e) -4

13. Calculate the following quantity:

$$\frac{d^2}{dt^2}\left(\frac{1}{1+t}\right)\bigg|_{t=1} =$$

(a) $\dfrac{1}{4}$

(b) 0

(c) $\dfrac{1}{2}$

(d) $-\dfrac{1}{4}$

(e) 1

14. Describe the way the slope of the graph changes as you move along the graph of the function given below.

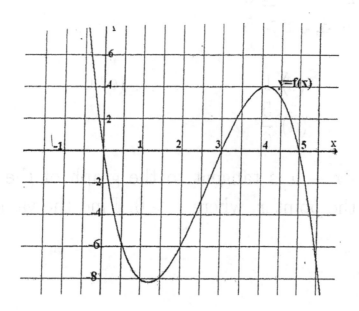

(a) The slope increases for $0.5 \leq x \leq 2$

(b) The slope decreases for $0.5 \leq x \leq 2$

(c) The slope increases for $3.5 \leq x \leq 5$

(d) The slope decreases for $0 \leq x \leq 1$

(e) The slope is not changing for $1 \leq x \leq 2$

15. Let $s(t) = -16t^2 + 128t + 20$ be the height of a ball (in feet), t seconds after it was thrown straight up into the air. At what time is the velocity -32 feet per second?

(a) $t = 5$

(b) $t = 896$

(c) $t = 3$

(d) $t = 20$

(e) $t = 160$

ANSWERS

EXAM 3

1. (A) ● (C) (D) (E)   11. (A) ● (C) (D) (E)
2. (A) (B) ● (D) (E)   12. (A) (B) (C) ● (E)
3. ● (B) (C) (D) (E)   13. ● (B) (C) (D) (E)
4. (A) (B) ● (D) (E)   14. ● (B) (C) (D) (E)
5. ● (B) (C) (D) (E)   15. ● (B) (C) (D) (E)
6. (A) (B) ● (D) (E)   16. (A) (B) (C) (D) (E)
7. (A) (B) (C) ● (E)   17. (A) (B) (C) (D) (E)
8. (A) (B) (C) ● (E)   18. (A) (B) (C) (D) (E)
9. (A) ● (C) (D) (E)   19. (A) (B) (C) (D) (E)
10. (A) ● (C) (D) (E)   20. (A) (B) (C) (D) (E)

1. Three hundred twenty dollars are available to fence in a rectangular garden. The fencing for the side of the garden facing the road costs $6.00 per foot and the other three sides cost $2.00 per foot. Find the dimensions of the largest possible such garden.

   (a) $x = 30$ ft and $y = 30$ ft   (d) $x = 40$ ft and $y = 20$ ft

   (b) $x = 25$ ft and $y = 75$ ft

   (c) $x = 40$ ft and $y = 20$ ft   (e) None of the above.

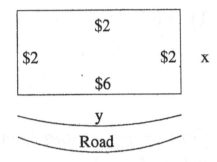

2. The cost function for a manufacturer is $C(x)$ dollars, where x is the number of units of goods produced. (See graph below).

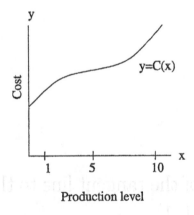

   Production level

   What is the economic significance of the inflection point?

   (a) The level of production has the least value.

   (b) The cost decreases as x increases.

   (c) The marginal cost has the least value.

   (d) The marginal cost has the greatest value.

   (e) The cost is greater than the profit.

3. The real number e is defined as the number for which the following statement is true:

(a) $\lim\limits_{h \to 0} (e^h + h) = 1$

(b) $\lim\limits_{h \to 0} \frac{e^h - 1}{h} = 1$

(c) $\lim\limits_{h \to 0} \frac{e^h - e^x}{h} = 1$

(d) $\lim\limits_{h \to 0} (e^h - 1) = 0$

(e) $\frac{d}{dx} e^x = 0$

4. The function $f(x) = \frac{x}{\ln x + x}$ does have a minimum point. Find the minimum value of $f(x)$.

(a) $\frac{1}{2}$

(b) 0

(c) $\frac{1}{\ln e}$

(d) e

(e) $\frac{e}{1+e}$

5. Find the slope of the tangent line to the curve $x^2 + 4xy + 4y^2 = 1$ at the point $(-1, 1)$

(a) 1

(b) 0

(c) $-\frac{1}{2}$

(d) -2

(e) none of the above.

6. Evaluate: $\displaystyle\lim_{h\to 0} \frac{e^{-(x+h)^2}-e^{-x^2}}{h}$

(a) xe^{x^2}

(b) $-xe^{x^2}$

(c) 1

(d) $-2xe^{x^2}$

(e) e^{-x^2}

7. The figure shows the graph of $f'(x)$, the derivative of a certain function $f(x)$. At $x = 4$, which of the following statements is true?

(a) $f(x)$ is increasing at $x = 4$.

(b) $f(x)$ is decreasing at $x = 4$.

(c) $f(x)$ has a relative minimum at $x = 4$.

(d) $f(x)$ has a relative maximum at $x = 4$.

(e) $f(x)$ is undefined at $x = 4$.

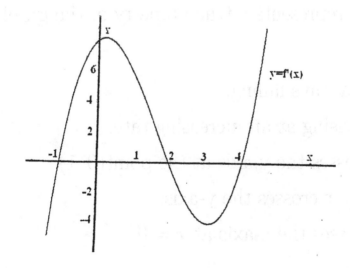

8. The height h of a cylinder is increasing, while the radius remains constant over a period of time. Find the rate of change of the total surface area of the cylinder.

(a) $\frac{dS}{dh} \cdot \frac{dS}{dt}$

(b) $2\pi r \frac{dh}{dt}$

(c) $4\pi r \frac{dr}{dt} + 2\pi h \frac{dr}{dt}$

(d) $2\pi r$

(e) 0

9. Let $f(x) = 3^x$. Determine $f'(x)$.

(a) $f'(x) = x \ln 3$

(b) $f'(x) = x + 3^x$

(c) $f'(x) = 3^{x-1}$

(d) $f'(x) = 3^x \ln 3$

(e) $f'(x) = \ln 3 + 3^x$

10. Which statement represents a true property of the graph of $y = \ln x$?

(a) The graph is always falling.

(b) The graph is rising at an increasing rate.

(c) The graph crosses the y-axis at the point$(0, 1)$

(d) The graph never crosses the y-axis.

(e) The graph crosses the x-axis at $x = 0$.

11. The revenue for a manufacturer is $R(x)$ thousand dollars, where x is the number of units of goods produced and sold. $R'(x)$ is the function given in the figure below. At what level of production is the revenue greatest?

(a) 40

(b) 35

(c) 0

(d) 42.5

(e) revenue is always falling

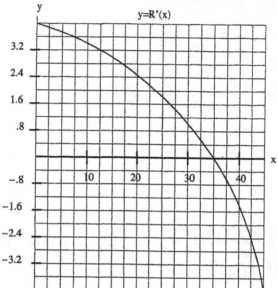

12. Find a formula for $\frac{d}{dx} f(g(x))$ Where $f(x)$ is a function such that $f'(x) = x\sqrt{1 - x^2}$ and $g(x) = \sqrt{x}$.

(a) $\frac{1}{2}\sqrt{1 - x}$

(b) $x\sqrt{1 - x^2} \cdot \frac{1}{2}x^{-\frac{1}{2}}$

(c) $\frac{x\sqrt{1-x^2}}{2\sqrt{x}}$

(d) $x\sqrt{x}\sqrt{1 - x^2}$

(e) $\frac{1}{2}\sqrt{1 - x} \cdot x^{\frac{1}{2}}$

13. Suppose that P, y, w, and h are variables, where P is a function of y and y is a function of w and w is a function of h. Write the chain rule for $\frac{dP}{dh}$

(a) $\frac{dP}{dh} = \frac{dP}{dy} \cdot \frac{dy}{dw} \cdot \frac{dw}{dh}$

(b) $\frac{dP}{dh} = \frac{dy}{dw} \cdot \frac{dP}{dy}$

(c) $\frac{dP}{dh} = \frac{dP}{dw} \cdot \frac{dw}{dh} \cdot \frac{dy}{dh}$

(d) $\frac{dP}{dh} = \frac{dP}{dw} \cdot \frac{dw}{dh} \cdot \frac{dP}{dh}$

(e) $\frac{dP}{dh} = \frac{dP}{dy} \cdot \frac{dy}{dw} \cdot \frac{dw}{dy}$

14. If $\frac{d^2 f(x)}{dx^2} = -12x^2$, then what can you conclude about the graph of $f(x)$?

(a) $f(x)$ has an inflection point at $x = 0$.

(b) $f(x)$ is always decreasing.

(c) $f(x)$ is always concave up.

(d) $f(x)$ is always increasing.

(e) $f(x)$ has no inflection points.

15. Find the point on the graph of $y = \frac{e^x}{x + e^x}$ where the tangent line is horizontal.

(a) $(0, 1)$

(b) $(1, \frac{e}{1+e})$

(c) $(-1, \frac{1}{1-e}$

(d) $(e, \frac{e^e}{e+e^e})$

(e) none of the above.

ANSWERS

EXAM 4

1 (A) (B) (C) ● (E)   11 (A) ● (C) (D) (E)
2 (A) (B) ● (D) (E)   12 ● (B) (C) (D) (E)
3 (A) ● (C) (D) (E)   13 ● (B) (C) (D) (E)
4 (A) (B) (C) (D) ●   14 (A) (B) (C) (D) ●
5 (A) (B) ● (D) (E)   15 (A) ● (C) (D) (E)
6 (A) (B) (C) ● (E)   16 (A) (B) (C) (D) (E)
7 (A) (B) ● (D) (E)   17 (A) (B) (C) (D) (E)
8 (A) ● (C) (D) (E)   18 (A) (B) (C) (D) (E)
9 (A) (B) (C) ● (E)   19 (A) (B) (C) (D) (E)
10 (A) (B) (C) ● (E)   20 (A) (B) (C) (D) (E)

167

1. The concentration C of a particular blood pressure medication t hours after it is swallowed is given by $C(t) = \dfrac{0.3t}{(t+2.5)^2}$. How long will it take for the concentration to be at its highest?

   (a) 1.0 hours

   (b) 2.0 hours

   (c) 1.5 hours

   (d) 2.5 hours

   (e) 3.0 hours

2. Find the slope of the tangent line to the curve $y = e^{x^2 - 5x + 10}$ at $(0, e^{10})$.

   (a) $-5e^{10}$

   (b) $-e^{10}$

   (c) $5e^{10}$

   (d) 1

   (e) $-5e^{-1}$

3. Which of the following statements is true?

   (a) If $f(x) = e^{22}$, then $f'(x) = e^{22}$

   (b) If $f(x) = \ln 12$, then $f'(x) = \frac{1}{12}$

   (c) If $f(x) = \ln a^x$, then $f'(x) = \ln x$

   (d) If $f(x) = \ln a^x$, then $f'(x) = \ln a$

   (e) If $f(x) = 5^x$, then $f'(x) = 5^x$

4. Let $f'(x) = (1 - x)e^{-x}$. What can you say about $f(x)$:

    (a) $f(x)$ has a relative minimum at $x = 1$

    (b) $f(x)$ has a relative maximum at $x = 0$

    (c) $f(x)$ is always rising

    (d) $f(x)$ is falling on the internal $(1, \infty)$

    (e) $f(x)$ has no relative extreme points

5. Find the limit

$$\lim_{h \to 0} \frac{\ln(e + h) - 1}{h}$$

    (a)   0   (d) 1

    (b) $\dfrac{d}{dx}(\ln e)$ (e) $\dfrac{1}{e}$

    (c) $\ln e$

6. Let $f(x) = x + \dfrac{4}{x}$, $f'(x) = 1 - \dfrac{4}{x^2}$, and $f''(x) = \dfrac{8}{x^3}$. Then

    (a) $f(x)$ has an oblique asymptote at $x = 4$

    (b) $f(x)$ has no relative extreme points

    (c) $f(x)$ has an inflection point at $x = 0$

    (d) $f(x)$ is always rising

    (d) $f(x)$ has no inflection points

7. A closed rectangular box with a square base and a volume of 96 cm^3 is to be constructed using two different types of materials. The top is made of metal costing $2 per square centimeter and the remainder is made of wood costing $1 per square centimeter. What is the minimum cost of constructing such a box?

   (a) $180

   (b) $132

   (c) $144

   (d) $180

   (e) $14.40

8. Let $f(x) = x^2 e^{-x}$. Find the coordinates of the relative maximum.

   (a) $(2, 4e^2)$

   (b) $(-2, \frac{-4}{e^2})$

   (c) $(-2, 4e^2)$

   (d) $(2, \frac{4}{e^2})$

   (e) None of the above.

9. Find the inflection point(s) of the function $f(x) = \ln(x^2 - 4)$.

   (a) $(\sqrt{5}, 0)$

   (b) $(-2, 0)$

   (c) $(2, 0)$

   (d) $(0, 0)$

   (e) There is no inflection point.

10. Find the greatest area of a rectangular garden that can be fenced on all four sides with 300 meters of fence.

    (a) 90,000 m^2

    (b) 80 m

    (c) 5.625 m^2

    (d) 75 m^2

    (e) 85 m

11. Let $f'(x) = \dfrac{-5}{(3x-1)^2}$. Then $f(x) = \dfrac{2x+1}{3x-1}$

    (a) is always rising

    (b) $f(x)$ has an inflection point at $x = 30$

    (c) $f(x)$ has a relative minimum at $x = -5$

    (d) $f(x)$ does not have any relative extreme points

    (e) $f(x)$ has a horizontal asymptote at $x = -5$

12. Evaluation: $\displaystyle\lim_{h \to 0} \dfrac{e^{2(x+h)^3} - e^{2x^3}}{h}$

    (a) $3e^8$

    (b) 0

    (c) It does not exist

    (d) $2x^3 e^{2x^3 - 1}$

    (e) $6x^2 e^{2x^3}$

13. Suppose W, x, b, c and t are variables such that W is a function of x, x is a function of b, b is a function of c, and c is a function of t. If W represents a profit function and t represents time, find the time rate of change of the profit function.

(a) $\dfrac{dW}{dx} \cdot \dfrac{dx}{db} \cdot \dfrac{db}{dt} \cdot \dfrac{dc}{db}$

(b) $\dfrac{dW}{dx} \cdot \dfrac{dx}{db} \cdot \dfrac{db}{dc} \cdot \dfrac{dc}{dt}$

(c) $\dfrac{dW}{dc} \cdot \dfrac{dc}{dt} \cdot \dfrac{dt}{db}$

(d) $\dfrac{dW}{dx} \cdot \dfrac{dx}{db} \cdot \dfrac{dc}{dt}$

(e) $\dfrac{dW}{dc} \cdot \dfrac{dx}{db}$

14 If $f(x) = \ln(x - 3)$, at what value of x is the slope of the tangent line to $f(x)$ equal to 2?

(a) 3.5

(b) 7

(c) 2.5

(d) 5

(e) 4.5

15. Let $f(x) = 2 + e^{4x}$. Find the limit of $f(x)$ as x approaches negative infinity, that is $\lim\limits_{x \to -\infty} f(x)$

(a) 0

(d) 2

(b) It does not exist.

(e) 4

(c) 1

ANSWERS

EXAM 5

1 Ⓐ Ⓑ Ⓒ ● Ⓔ   11 Ⓐ Ⓑ Ⓒ ● Ⓔ
2 ● Ⓑ Ⓒ Ⓓ Ⓔ   12 Ⓐ Ⓑ Ⓒ Ⓓ ●
3 Ⓐ Ⓑ Ⓒ ● Ⓔ   13 Ⓐ ● Ⓒ Ⓓ Ⓔ
4 Ⓐ Ⓑ Ⓒ ● Ⓔ   14 ● Ⓑ Ⓒ Ⓓ Ⓔ
5 Ⓐ Ⓑ Ⓒ Ⓓ ●   15 Ⓐ Ⓑ Ⓒ ● Ⓔ
6 Ⓐ Ⓑ Ⓒ Ⓓ ●   16 Ⓐ Ⓑ Ⓒ Ⓓ Ⓔ
7 Ⓐ Ⓑ ● Ⓓ Ⓔ   17 Ⓐ Ⓑ Ⓒ Ⓓ Ⓔ
8 Ⓐ Ⓑ Ⓒ ● Ⓔ   18 Ⓐ Ⓑ Ⓒ Ⓓ Ⓔ
9 Ⓐ Ⓑ Ⓒ Ⓓ ●   19 Ⓐ Ⓑ Ⓒ Ⓓ Ⓔ
10 Ⓐ Ⓑ ● Ⓓ Ⓔ   20 Ⓐ Ⓑ Ⓒ Ⓓ Ⓔ

EXAM 6

1. An open-top rectangular box with a square base x by x and height h is going to be constructed out of cardboard. The volume of the box has to be 62.5 cubic feet. Find the height h of the box that will yield the minimum surface area.

   (a) 10 ft
   (b) 5 ft
   (c) 2.5 ft
   (d) 125 ft
   (e) 7.5 ft

2. Determine the slope of the graph:

$$(xy^3 + y)^4 = 16$$

   at the point $(1,1)$.

   (a) 2
   (b) -2
   (c) $\frac{1}{4}$
   (d) $-\frac{1}{4}$
   (e) 0

3. Determine whether the following function is continuous and/or differentiable at $x = 0$, where $f(x)$ is given by

$$f(x) = \begin{cases} 4 - 2x, & \text{for} \quad x < 0 \\ 4, & \text{for} \quad x = 0 \\ 3 + e^x, & \text{for} \quad x > 0 \end{cases}$$

   (a) f is continuous and differentiable at $x = 0$
   (b) f is continuous, but not differentiable at $x = 0$
   (c) f is not continuous and not differentiable at $x = 0$
   (d) f is not continuous, but is differentiable at $x = 0$
   (e) None of the above.

4. Evaluate: $\lim\limits_{h\to 0} \dfrac{e^{-(x+h)^2} - e^{-x^2}}{h}$

(a) xe^{x^2} (b) $-xe^{x^2}$ (c) 1   (d) $-2xe^{-x^2}$ (e) e^{-x^2}

5. Which of the following four properties are true of the graph of $y = 10e^{2x}$?

   1. It is concave up.

   2. The $y-$ intercept is $(0, 2)$.

   3. It has a minimum at $x = 0$.

   4. y is positive for $x > 0$ and negative for $x < 0$.

(a) **(3)** and **(4)**   (c) **(1)** only   (e) **(4)** only

(b) **(2)** and **(4)**   (d) **(2)** and **(3)**

6. Find the values of x at which the function $f(x) = e^{-2x} + 2x$ has a possible maximum or minimum point.

(a) Maximum at $x = \frac{0.69}{2}$

(b) Maximum at $x = \frac{e}{2}$

(c) Minimum at $x = 0$

(d) There are no relative maximum or minimum points

(e) None of the above

7. Some years ago it was estimated that the demand for steel approximately satisfied the equation $p = 151 - 25x$, and the total cost of producing x units of steel was $C(x) = 179 + 51x$. The quantity x was measured in millions tons and the price and total cost were measured in millions of dollars. Determine the price that maximizes the profits.

(a) 1   (d) 101

(b) 2   (e) 230

(c) 156

8. Find the slope of the tangent line to the graph of $y = 2x + \ln\left(\frac{1}{x}\right)$ at $x = 1$

(a) 0   (d) 3

(b) 1   (e) 4

(c) 2

9. Let $f(x)$ and $g(x)$ be differentiable functions where $g(x) = \sqrt[3]{x}$. Find an expression for $\frac{d}{dx} f(g(x))$.

(a) $\sqrt[3]{x} f'(x) + \frac{1}{3} x^{-\frac{2}{3}} f(x)$

(b) $\frac{1}{3} x^{-\frac{2}{3}} \cdot g'(x)$

(c) $\frac{1}{3\sqrt[3]{x^2}} f'(\sqrt[3]{x})$

(d) $\sqrt[3]{x} f(\sqrt[3]{x})$

(e) None of the above.

10. Let $f(x)$ be a function with its second derivative given by $f''(x) = (x+1)^2$. What is the inflection point on the graph of $f(x)$?

(a) $x = -1$

(b) $x = 0$

(c) $x = 1$

(d) f has no inflection point

(e) None of the above

11. The function $f(x)$ has its first derivative equal to $\dfrac{2x-2}{x^2 - 2x + 5}$. Which of the following statements is true?

(a) $f(x)$ is always increasing

(b) $f(x)$ has a local max

(c) $f(x)$ is increasing for $x > 1$

(d) $f(x)$ is always decreasing

(e) $f(x)$ is increasing for $x < 1$

12. Find $\dfrac{du}{dx}$, where y is a function of u such that $\dfrac{dy}{du} = \dfrac{u}{u^2 + 1}$ and $u = x^{\frac{3}{2}}$

(a) $\dfrac{du}{dx} = \dfrac{1-u^2}{(u^2+1)^2} \cdot \dfrac{3}{2} x^{\frac{1}{2}}$

(b) $\dfrac{du}{dx} = \dfrac{u}{u^2+1} x^{\frac{3}{2}}$

(c) $\dfrac{du}{dx} = \dfrac{x^{\frac{3}{2}}}{x^{\frac{3}{2}}+1} \cdot \dfrac{3}{2} x^{\frac{5}{2}}$

(d) $\dfrac{du}{dx} = \dfrac{3x^2}{2(x^3+1)}$

(e) $\dfrac{du}{dx} = \dfrac{3}{2} \cdot \sqrt{x}$

13. Evaluate $\lim\limits_{h \to 0} \dfrac{\ln(7+h) - \ln 7}{h}$

(a) $\frac{d}{dx} \ln 7$

(b) 0

(c) $\frac{1}{7}$

(d) $\frac{1}{h}$

(e) $\frac{1}{10}$

14. Let

$$f(x) = \begin{cases} x^3 & \text{if} \quad 0 \le x < 1 \\ 2 & \text{if} \quad x = 1 \\ \ln x & \text{if} \quad 1 < x < 5 \end{cases}$$

Find the $\lim\limits_{x \to 1^+} f(x)$.

(a) 1

(b) 0

(c) does not exist

(d) 2

(e) 5

15. Find the equation of the tangent line to the curve $y = \dfrac{e^x}{x + e^x}$ at the point $(0, 1)$.

(a) $y = 1$

(b) $y = x - 1$

(c) $y = -x + 1$

(d) $\frac{1}{2}y = -x + \frac{1}{2}$

(e) $y = \frac{e}{1+e}$

ANSWERS

EXAM 6

1	Ⓐ	Ⓑ	●	Ⓓ	Ⓔ	11	Ⓐ	Ⓑ	●	Ⓓ	Ⓔ
2	Ⓐ	Ⓑ	Ⓒ	●	Ⓔ	12	Ⓐ	Ⓑ	Ⓒ	Ⓓ	●
3	Ⓐ	●	Ⓒ	Ⓓ	Ⓔ	13	Ⓐ	Ⓑ	●	Ⓓ	Ⓔ
4	Ⓐ	Ⓑ	Ⓒ	●	Ⓔ	14	Ⓐ	●	Ⓒ	Ⓓ	Ⓔ
5	Ⓐ	Ⓑ	●	Ⓓ	Ⓔ	15	Ⓐ	Ⓑ	●	Ⓓ	Ⓔ
6	Ⓐ	Ⓑ	●	Ⓓ	Ⓔ	16	Ⓐ	Ⓑ	Ⓒ	Ⓓ	Ⓔ
7	Ⓐ	Ⓑ	Ⓒ	●	Ⓔ	17	Ⓐ	Ⓑ	Ⓒ	Ⓓ	Ⓔ
8	Ⓐ	●	Ⓒ	Ⓓ	Ⓔ	18	Ⓐ	Ⓑ	Ⓒ	Ⓓ	Ⓔ
9	Ⓐ	Ⓑ	●	Ⓓ	Ⓔ	19	Ⓐ	Ⓑ	Ⓒ	Ⓓ	Ⓔ
10	Ⓐ	Ⓑ	Ⓒ	●	Ⓔ	20	Ⓐ	Ⓑ	Ⓒ	Ⓓ	Ⓔ

1. The figure below shows the graphs of several functions $f(x)$ for which $f'(x) = \dfrac{2}{x}$. Find the expression for the function $f(x)$ whose graph passes through the point $(1, 2)$.

   (a) $2x$

   (b) $2\ln|x| + 2$

   (c) $\ln|x| + 1$

   (d) $\dfrac{2}{x^2}$

   (e) $\ln|x|$

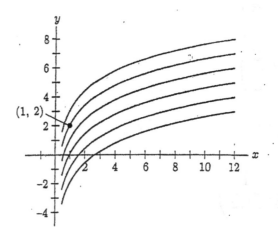

2. Let

$$f(x) = \begin{cases} 3x + 4, & x \le -1 \\ |x|, & x > -1 \end{cases}$$

   At $x = -1$, the function $f(x)$ is:

   (a) Not continuous and not differentiable.

   (b) Continuous and differentiable.

   (c) Continuous but not differentiable.

   (d) Not continuous but differentiable.

   (e) None of the above.

3. The relative maximum point of $y = x \cdot e^{-x}$ is:

(a) $(-1, e)$

(b) $\left(1, \dfrac{1}{e}\right)$

(c) $\left(2, \dfrac{2}{e}\right)$

(d) $(0, 0)$

(e) $\left(\dfrac{1}{2}, \dfrac{1}{e^{1/2}}\right)$

4. The velocity of a truck (in feet per second) t seconds after starting from rest is given by the function $V(t) = 12\sqrt[3]{t}, (0 \leq t \leq 10)$. Find $s(t)$, the function that give the position of the truck at anytime t.

(a) $3t^{\frac{4}{3}}$

(b) $9t^{\frac{4}{3}}$

(c) $4t^{\frac{-1}{3}}$

(d) $16t^{\frac{4}{3}}$

(e) None of the above.

5. Given two points $P = (a, f(a))$ and $Q = (b, f(b))$ on the graph of the function $y = f(x)$ (as the picture shows), find the slope of the secant line from P to Q.

(a) $f(x + h) - f(x)$

(b) $\dfrac{f(b) - f(a)}{b - a}$

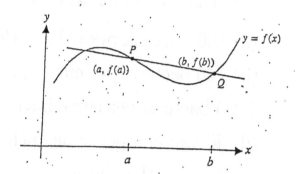

(c) $\dfrac{b - a}{b}$

(d) $f'(x)$

(e) $\lim\limits_{x \to 0}[f(x + h) - f(x)]$

6. The maximum **value** of the function $f(x) = x^3 - 12x + 8$ for $x \leq 0$ is:

(a) 24

(b) -8

(c) -2

(d) 40

(e) None of the above.

7. Determine $g(x)$ where $\displaystyle\int g(x)dx = \dfrac{d}{dx}(e^{3x} - 2x^3) + C$

(a) $g(x) = e^{3x} - 2x^3$
(b) $g(x) = \dfrac{e^{3x}}{3} - \dfrac{x^4}{2}$
(c) $g(x) = 9e^{3x} - 12x$

(d) $g(x) = \dfrac{d}{dx}(e^{3x} - 2x^3)$

(e) None of the above.

8. A small body of water is treated to control bacterial growth. After t days, the concentration of bacteria per cubic inch $C(t)$ is found to be:

$$C(t) = 20t^2 - 200t + 640$$

What is the lowest bacterial concentration?

(a) 500 bacteria per cubic inch

(b) 140 bacteria per cubic inch

(c) 5 bacteria per cubic inch

(d) 40 bacteria per cubic inch

(e) None of the above

9. A rock thrown straight up in the air has a velocity of $v(t) = -9.8t + 20$ meters per second after t seconds. Assume the rock starts at time $t = 0$ on the ground ($s(0) = 0$). The distance the rock travels during the first 2 seconds is best represented by the shaded area in which of the following graphs?

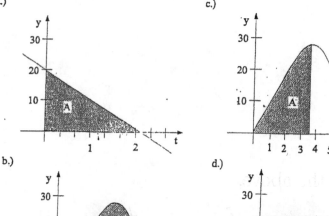

(a) Graph a).

(b) Graph b).

(c) Graph c).

(d) Graph d).

(e) None of the above.

10. Suppose $p(t)$ is the rate (in tons per year) at which pollutants are discharged into a lake, where t is the number of years since 1990.

Interpret $\int_5^7 p(t)\, dt$.

(a) The number of tons of pollutants discharged from 1995 to 1997.

(b) The rate of pollutants in tons per year for the year 1995 and the year 1997.

(c) $\frac{dp(t)}{dt}$ tons/year.

(d) The total number of years.

(e) The total number of tons of pollutants from 1990 to today.

11. Given that $\frac{d}{dx}(x\ln x)=\ln x + 1$, find the value of the integral

$$\int_1^2 (\ln x + 1)\, dx$$

(a) $\ln x$

(b) $\left(\frac{\ln x}{2}\right)^2 + x$

(c) $2\ln 2$

(d) $\dfrac{\ln 2}{2}$

(e) $\ln 1 - \ln 2$

12. Find the area of the region bounded by the parabola

$$y = -x^2 + 5 \text{ and the line } y = 1$$

(a) $\frac{30}{2}$

(b) 0

(c) $-\frac{32}{2}$

(d) $\frac{32}{3}$

(e) $\frac{30}{3}$

13. A rock is thrown off a cliff. At t seconds, its distance from the ground below is $s(t) = -16t^2 - 16t + 96$ feet. How fast will the rock be falling when it hits the ground?

(a) -32 ft/sec

(b) -80 ft/sec

(c) -16 ft/sec

(d) 0 ft/sec

(e) $-32t$

14. Evaluate: $\lim\limits_{x \to 1} \left(\ln x + \dfrac{3x^2 - 3}{x - 1} \right)$

(a) 3

(b) e

(c) It does not exist.

(d) 2

(e) 6.

15. Suppose the revenue from selling x custom-made office desks is:

$$R(x) = 2000\left(1 - \frac{1}{x+1}\right) \text{ dollars.}$$

Find the marginal revenue when x desks are sold.

(a) $\frac{2000}{(x+1)^2}$

(b) $-2000(x+1)^{-1}$

(c) ∞

(d) It does not exist.

(e) $2000 - \frac{2000}{(x+1)^2}$

16. Cars A and B start at the same place and travel in the same direction, with velocities after t hours given by the functions $v_A(t)$ and $v_B(t)$ respectively as represented in the figure below. What does the area between the two curves from $t = 0$ to $t = 1$ represent?

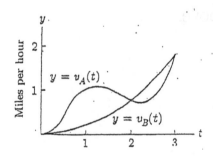

a) The total distance covered in one hour.

b) The final velocity.

c) The distance between the two cars after 1 hour.

d) The maximum velocity.

e) The distance covered minus the velocity.

17. Which of the following properties are true of the graph of
$y = 10e^{2x}$?

(a) It is concave up.

(b) The y-intercept is $(0, 2)$.

(c) It has a minimum at $x = 0$.

(d) y is positive for $x > 0$ and negative for $x < 0$.

(e) None of the above.

18. Find the relative extrema for the function

$$f(x) = x^3 - 3x^2$$

(a) Relative maximum at $(0, 0)$; Relative minimum at $(2, -4)$

(b) Relative maximum at $(2, -4)$; Relative minimum at $(0, 0)$

(c) Relative maximum at $(-2, -20)$; Relative minimum at $(2, -4)$

(d) Relative maximum at $(2, -4)$; Relative minimum at $(-2, -20)$

(e) None of the above.

19. Evaluate: $\int_{-1}^{1} e^{-2x}\, dx$

(a) $e^2 - e^{-2}$

(b) $\frac{1}{2}(e^{-2} - e^2)$

(c) $\frac{1}{2}(e^2 - e^{-2})$

(d) $e^{2x} + k$

(e) $-\frac{1}{2}e^{-2x} + c$

20. A ball is thrown into the air. Let $s(t)$ be the height of the ball in feet after t seconds. Suppose $s(t) = -(4t - 9)^2$. What is the acceleration of the ball after 5 seconds?

(a) 0 ft/sec^2

(b) -32 ft/sec^2

(c) 16 ft/sec^2

(d) -88 ft/sec^2

(e) None of the above

21. Consider the following figure. If $MC(x)$ is the marginal cost function for producing x units of a certain product and $MR(x)$ is the corresponding marginal revenue function, then what does the shaded area represent?

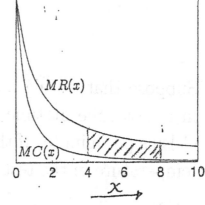

(a) the total revenue for producing 4 units

(b) the cost plus the profit for producing 4 units

(c) the total revenue plus the total cost for producing 4 units

(d) the profit between production level $x = 4$ and $x = 8$

(e) None of the above

22. A flu epidemic hits a town. Public Health officials estimate that the number of persons sick with the flu at the time t (measured in days from the beginning of the epidemic) is approximated by $P(t) = 60t^2 - t^3$, provided that $0 \leq t \leq 40$. When is the flu spreading at the rate of 900 people per day?

(a) Never

(b) $t = 60$ days

(c) $t = 10$ days and $t = 30$ days

(d) $t = 20$ days

(e) None of the above.

23. Suppose that during a controlled experiment, the temperature in a test tube at time t is rising at a rate of $6t^2 + 2$ degrees Celsius per minute. If the initial temperature is $0°C$, what is the temperature in the test tube after 5 minutes?

(a) 28

(b) 260

(c) 152

(d) 760

(e) None of the above.

24. Write down a definite integral or sum of definite integrals that gives the area of the shaded portion of the figure below.

(a) $\displaystyle\int_{1}^{4} f(x)\,dx$

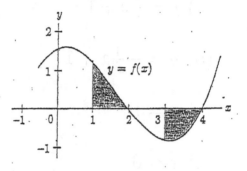

(b) $\displaystyle\int_{0}^{4} f(x)\,dx$

(c) $\displaystyle\int_{1}^{2} f(x)\,dx + \int_{3}^{4} -f(x)\,dx$

(d) $\displaystyle\int_{1}^{2} [f(x) + g(x)]\,dx + \int_{3}^{4} [f(x) - g(x)]\,dx$

(e) $\displaystyle\int_{1}^{2} [f(x) - g(x)]\,dx$

25. If $f(x) = \dfrac{1}{1+x}$, then $f'(2)$ is equal to:

(a) $\displaystyle\lim_{x \to 2} \dfrac{1}{1+x}$

(b) $\displaystyle\lim_{h \to 0} \dfrac{\frac{1}{3+h} - \frac{1}{3}}{h}$

(c) $\dfrac{1}{3}$

(d) $-\dfrac{1}{3(3+h)}$

(e) None of the above.

26. Write the equation of the tangent line to the graph of the function $f(x) = \ln(x^2 + x + e)$ at the point $x = 0$.

(a) $y = \frac{1}{e}x + 1$

(b) $y = -\frac{1}{e^2}x + 1$

(c) $y = 1$

(d) $x = 0$

(e) None of the above.

27. Interpret the area of the shaded region in the figure below. The function in the graph represents the rate of soil erosion.

(a) rate of soil erosion

(b) $\frac{df(t)}{dt}$

(c) marginal erosion

(d) $\int_0^5 dt$

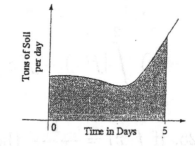

(e) tons of soil eroded during a five day period.

28. Evaluate: $\int \dfrac{1 + x^2 e^{-2x}}{x^2} dx$

(a) $\dfrac{1}{x^3} - \dfrac{1}{2}e^{-2x} + C$

(c) $x^{-2} - \dfrac{1}{2}e^{-2x} + C$

(b) $-\dfrac{1}{x} - \dfrac{1}{2e^{2x}} + C$

(d) $\dfrac{1}{x} + \dfrac{1}{2}e^{2x} + C$

(e) None of the above.

29. Let $f(x)$ and $g(x)$ be functions, where $g(x) = \sqrt[3]{x}$. Use the chain rule to find an expression for $\dfrac{d}{dx}f(g(x))$.

(a) $\sqrt[3]{x}f'(x) + \dfrac{1}{3}x^{-\frac{2}{3}}f(x)$

(b) $\dfrac{1}{3}x^{-\frac{2}{3}} \cdot g'(x)$

(c) $\dfrac{1}{3\sqrt[3]{x^2}} \cdot f'(\sqrt[3]{x})$

(d) $\sqrt[3]{x} \cdot f(\sqrt[3]{x})$

(e) None of the above.

30. Suppose P, y, x and t are variables such that P is a function of y, y is a function of x, and x is a function of t. If P represents a profit function and t represents time, find the time rate of change of the profit function.

(a) $\dfrac{dP}{dy} \cdot \dfrac{dy}{dt}$

(b) $\dfrac{dP}{dx} \cdot \dfrac{dt}{dx}$

(c) $\dfrac{dP}{dy} \cdot \dfrac{dx}{dt}$

(d) $\dfrac{dP}{dy} \cdot \dfrac{dy}{dx} \cdot \dfrac{dx}{dt}$

(e) None of the above.

ANSWERS

EXAM 7

1 (A) ● (C) (D) (E)	11 (A) (B) ● (D) (E)	21 (A) (B) (C) ● (E)
2 (A) (B) ● (D) (E)	12 (A) (B) (C) ● (E)	22 (A) (B) ● (D) (E)
3 (A) ● (C) (D) (E)	13 (A) ● (C) (D) (E)	23 (A) ● (C) (D) (E)
4 (A) ● (C) (D) (E)	14 (A) (B) (C) (D) ●	24 (A) (B) ● (D) (E)
5 (A) ● (C) (D) (E)	15 ● (B) (C) (D) (E)	25 (A) ● (C) (D) (E)
6 ● (B) (C) (D) (E)	16 (A) (B) ● (D) (E)	26 ● (B) (C) (D) (E)
7 (A) (B) ● (D) (E)	17 ● (B) (C) (D) (E)	27 (A) (B) (C) (D) ●
8 (A) ● (C) (D) (E)	18 ● (B) (C) (D) (E)	28 (A) ● (C) (D) (E)
9 ● (B) (C) (D) (E)	19 (A) (B) ● (D) (E)	29 (A) (B) ● (D) (E)
10 ● (B) (C) (D) (E)	20 (A) ● (C) (D) (E)	30 (A) (B) (C) ● (E)

1. Find $\dfrac{dy}{dx}$ where y is a function of u such that

$$\frac{dy}{du} = \frac{u}{u^2 + 1} \text{ and } u = x^{\frac{3}{2}}.$$

(a) $\dfrac{dy}{dx} = \dfrac{1 - u^2}{(u^2 + 1)^2} \cdot \dfrac{3}{2}x^{\frac{1}{2}}$

(d) $\dfrac{dy}{dx} = \dfrac{3x^2}{2(x^3 + 1)}$

(b) $\dfrac{dy}{dx} = \dfrac{u}{u^2 + 1} \cdot x^{\frac{2}{3}}$

(e) $\dfrac{dy}{dx} = \dfrac{3}{2} \cdot \sqrt{x}$

(c) $\dfrac{dy}{dx} = \dfrac{x^{\frac{3}{2}}}{x^{\frac{3}{2}} + 1} \cdot \dfrac{3}{2}x^{\frac{5}{2}}$

2. The minimum value of the slope of the tangent line to the curve of $f(x) = x^5 + x^3 - 2x$ is:

(a) 0

(d) Does not exist.

(b) -2

(e) None of the above.

(c) 1

3. Let $f(1) = 3$, $f'(1) = 0$ and $f''(1) = 6$. Then $(1, 3)$ must be:

(a) relative maximum point.

(d) absolute minimum point.

(b) relative minimum point.

(e) point of discontinuity.

(c) point of inflection.

4. The graph of $y = (\ln x)^2$ has an inflection point at:

(a) $(e, 1)$

(b) $(1, 0)$

(c) (e, e^2)

(d) $(1, e^2)$

(e) None of the above.

5. Determine on which interval(s) the graph of $f(x) = \dfrac{e^x}{x^2}$ is rising.

(a) $(0, 2)$

(b) $(-\infty, 0), (2, \infty)$

(c) $(-\infty, 2)$

(d) $(2, -\infty)$

(e) None of the above

6. A small company has determined that the daily output y of a product t days after the start of production is $y = 150 - 150e^{-0.2t}$. At what rate is the daily output increasing when $t = 10$ days? ($Note : e \approx 2.7$).

(a) 11.11

(b) 120

(c) 1

(d) 4.12

(e) 1.45

7. Given: $\frac{d}{dx}(-e^{-x} - xe^{-x}) = xe^{-x}$ the value of $\int_{0}^{1} xe^{-x}dx$ is:

(a) $1 - \frac{2}{e}$

(c) $-2e^{-1}$

(d) $-e^{-1} + 1$

(b) 0

(e) e^{-1}

8. Suppose that a function $f(x)$ has the following properties:
$f'(x)$ is integrable on $[1, 3]$ and $f(1) = 0$.

Simplify the expression $\int_{1}^{3} f'(x)dx$.

(a) $f'(3) - f'(1)$

(d) $f(1)$

(b) $f'(3)$

(e) $-f(1)$

(c) $f(3)$

9. Suppose that the velocity of an object at anytime t measured in feet per second is given by $v(t) = -32t + 32$. Find its position $s(t)$, at any time t given that the initial position $(t = 0)$ was 200 feet above the ground.

(a) $-16t^2 + 32t + 200$

(d) $-32t^2 + 32 + 200$

(b) 32 feet

(e) $\frac{t^2}{2} - 200$

(c) $-16t^2 + 32 - 200$

10. Let $f(t) = -\frac{1}{5t^5}$. Find the instantaneous rate of change of $f(t)$ at $t = 1$.

(a) 1

(b) $\frac{dy}{dt}$

(c) $-t^{-5}$

(d) 6

(e) None of the above.

11. The function $f(x) = e^{-x^2}$ has point(s) of inflection at:

(a) $x = 0$

(b) $x = \pm 2$

(c) $x = \pm\frac{1}{2}$

(d) $x = \pm\frac{1}{\sqrt{2}}$

(e) Nowhere.

12. Let $f(x)$ be a function with its second derivative given by $f''(x) = (\ln(x+1))^2$. What is the inflection point on the graph of $f(x)$?

(a) $x = -1$

(b) $x = 0$

(c) $x = 1$

(d) f has no inflection point.

(e) None of the above.

13. Find the slope of the tangent line to the graph of $y = \ln(xe^x)$ at the point where $x = 3$.

(a) $\frac{4}{3}$

(b) $\ln 3 - 3$

(c) $e^3 + \ln 3$

(d) $\frac{1}{3}$

(e) None of the above

14. What is the area under the curve $y = \frac{1}{2x+1}$ from $x = 0$ to $x = 3$?

    (a) $\frac{1}{2}\ln 7$

    (b) $\ln 7 - \ln 1$

    (c) $-\frac{1}{7} + 1$

    (d) $\frac{1}{2}\ln(\frac{7}{2})$

    (e) None of the above.

15. $\displaystyle\int_0^2 (-3x^2 + 6x + 5)\,dx$ is equal to:

    (a) 5

    (b) 0

    (c) 14

    (d) 22

    (e) None of the above

16. What is the slope of the curve $y = \sqrt[3]{x^2 - 2x + 5}$ at $x = 1$?

    (a) $\frac{1}{12}$

    (b) $\frac{1}{3}$

    (c) $\frac{1}{6}$

    (d) $\frac{2}{3}$

    (e) None of the above.

17. Set up a definite integral or a sum of integrals that will give the area under the graph of the function $f(x)$ from $x = -3$ to $x = 4$, where $f(x) = \begin{cases} 9 \text{ if } x < 3 \\ x^2 \text{ if } x \geq 3 \end{cases}$

(a) $\displaystyle\int_{-3}^{4} x^2\,dx$

(b) $\displaystyle\int_{-3}^{3} 9\,dx + \int_{3}^{4} x^2\,dx$

(c) $\displaystyle\int_{-3}^{4} (x^2 - 9)\,dx$

(d) $\displaystyle\int_{-3}^{4} (x^2 + 9)\,dx$

(e) $\displaystyle\int_{-3}^{3} (9 - x^2)\,dx - \int_{3}^{4} (x^2 - 9)\,dx$

18. The figure below shows the graphs of several functions $f(x)$ for which $f'(x) = \frac{1}{3}$. Find the expression for the function $f(x)$ whose graph passes through (6,3).

(a) $f(x) = 0$

(b) $f(x) = 3x$

(c) $f(x) = \frac{1}{3}x + 1$

(d) $f(x) = \frac{x^2}{2} + \frac{1}{3}$

(e) $f(x) = x + \frac{1}{3}$

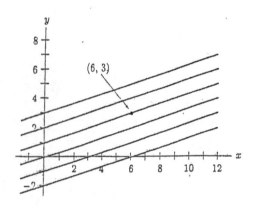

19. A farmer has 1200 yards of fencing material to build an E-shaped fence with which to enclose a rectangular piece of land along the straight side of a river. If fencing is not required along the river, and if he wants to create two identical rectangular pastures as in the picture, then find the largest area that he can enclose.

(a) $120,000 \ yd^2$

(b) $125,000 \ yd^2$

(c) $100,000 \ yd^2$

(d) $625,000 \ yd^2$

(e) $600,000 \ yd^2$

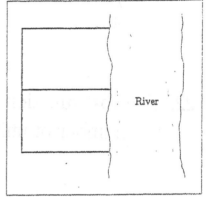

Figure   Rectangular pastures along a river.

20. Determine the value of c so that $f(x)$ is continuous on the entire set of real numbers when $f(x) = \begin{cases} x + 3, & x \le -1 \\ 2x - c, & x > -1 \end{cases}$

(a) -4

(b) 4

(c) 0

(d) -1

(e) None of the above

21. Which of the following represents the area bounded by $y = x^2 + 1$ and $y = -5x - 3$?

(a) $\displaystyle\int_{1}^{4} ((x^2 + 1) - (-5x - 3))\, dx$

(b) $\displaystyle\int_{-4}^{-1} ((-5x - 3) - (x^2 + 1))\, dx$

(c) $\displaystyle\int_{-4}^{-1} ((x^2 + 1) - (-5x - 3))\, dx$

(d) $\displaystyle\int_{1}^{4} ((-5x - 3) - (x^2 + 1))\, dx$

(e) None of the above

22. The cost function for a manufacturer is $C(x)$ dollars, where x is the number of units of goods produced. (See graph below)

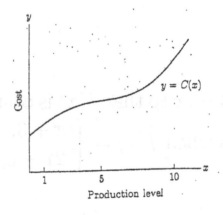

What is the economic significance of the inflection point?

(a) The level of production has the least value.

(b) The cost decreases as x increases.

(c) The marginal cost has the least value.

(d) The marginal cost has the greatest value.

(e) The cost is greater than the profit.

23. If $A(t)$ is the annual rate of world consumption of oil at time t (with $t = 0$ corresponding to the year 1987), which of the following expressions represents the amount of oil consumed between 1999 and 2009?

(a) $A'(10)$

(b) $\int_0^{10} A'(t) \, dt$

(c) $\int_0^{10} A(t) \, dt$

(d) $\int_{1999}^{2009} A(t) \, dt$

(e) $\int_{12}^{22} A(t) \, dt$

24. The function $g(x)$ in the figure below was obtained by shifting the graph of $f(x)$ up two units. If $h(x) = f(x) + 2$, then what is the derivative of $h(x)$?

(a) $g''(x)$

(b) $f'(x)$

(c) 0

(d) 2

(e) $\int [g(x) - f(x)] dx$

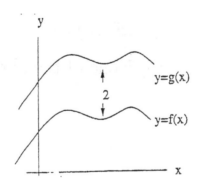

25. The cost of manufacturing x cases of cereal is C dollars, where $C = 3x + 4\sqrt{x} + 2$. Weekly production at t weeks from the present is estimated to be $x = 40 + 20t$ cases. How fast with respect to time are costs rising when $t = 3$?

(a) 20

(b) 80

(c) 64

(d) 5

(e) None of the above.

26. As h approaches 0, what value is approached by the difference quotient $\dfrac{e^h - 1}{h}$?

(a) 0

(b) 1

(c) e^x

(d) no value

(e) h

27. The tangent line to the graph of a function $f(x)$ at $(x, f(x))$ has slope $f'(x) = \frac{3}{x^2}$. Given that $(-1, 5)$ lies on the graph, determine $f(x)$.

(a) $3x^{-2} + C$

(b) $-6x^{-3}$

(c) $\frac{3}{x^3} + 2$

(d) $\frac{-3}{x} + 2$

(e) None of the above.

28. Let $f(x)$ and $g(x)$ be differentiable functions such that
$f(1) = 2, f'(1) = 3, g(1) = 4$ and $g'(1) = 5$.
Find $\dfrac{d}{dx}\left(\dfrac{f(x)}{g(x)}\right)\Big|_{x=1}$

(a) $\frac{1}{8}$

(d) $-\frac{3}{16}$

(b) 22

(e) $\frac{3}{5}$

(c) $\frac{3}{16}$

29. If the demand function for a monopolist is $p = 150 - .02x$ and
the cost function is $C(x) = 10x + 300$, find the value of x that
maximizes the profit.

(a) 3500

(d) 244,700

(b) 80

(e) None of the above.

(c) 35,300

30. In the figure below the straight line is tangent to the graph of
$f(x) = \frac{1}{x}$ at the point where $x = 2$. Find the value of a.

(a) 0

(b) 1

(c) 3

(d) 4

(e) $\frac{1}{3}$

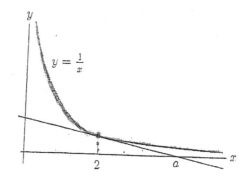

ANSWERS

EXAM 8

1 (A) (B) (C) ● (E) 11 (A) (B) (C) ● (E) 21 (A) ● (C) (D) (E)
2 (A) ● (C) (D) (E) 12 (A) (B) (C) ● (E) 22 (A) (B) ● (D) (E)
3 (A) ● (C) (D) (E) 13 ● (B) (C) (D) (E) 23 (A) (B) (C) (D) ●
4 ● (B) (C) (D) (E) 14 ● (B) (C) (D) (E) 24 (A) ● (C) (D) (E)
5 (A) ● (C) (D) (E) 15 (A) (B) ● (D) (E) 25 (A) (B) ● (D) (E)
6 (A) (B) (C) ● (E) 16 (A) (B) (C) (D) ● 26 (A) ● (C) (D) (E)
7 ● (B) (C) (D) (E) 17 (A) ● (C) (D) (E) 27 (A) (B) (C) ● (E)
8 (A) (B) ● (D) (E) 18 (A) (B) ● (D) (E) 28 ● (B) (C) (D) (E)
9 ● (B) (C) (D) (E) 19 ● (B) (C) (D) (E) 29 ● (B) (C) (D) (E)
10 ● (B) (C) (D) (E) 20 ● (B) (C) (D) (E) 30 (A) (B) (C) ● (E)